**IET POWER AND ENERGY SERIES 21**

Series Editors: Prof. A.T. Johns
J.R. Platts

T0258372

# Electricity Distribution Network Design

## 2nd Edition

## Other volumes in this series:

# Electricity Distribution Network Design

## 2nd Edition

E. Lakervi and E.J. Holmes

The Institution of Engineering and Technology

# Preface to 2nd edition

The six years since this book was first published have seen a number of developments in the area of distribution network design.

Market conditions are exerting stronger pressures on the design engineer, the structure of electrical utilities is changing or under review in a number of countries, and increasingly national or international legislation is adding further constraints. While many of the techniques of earlier decades are still in use, the increasing penetration of computers has significantly affected the manner in which distribution networks are being designed and operated.

As a consequence, Chapter 14, covering computer-based planning, has been re-written, as has Chapter 4 on reliability. Advantage has been taken to add material on international recommendations affecting network design and to discard some outdated and less relevant material. Also, a number of the revisions are based on students' comments arising when the first edition was being used for teaching international university and short courses.

Once again, we acknowledge the help of our friends and colleagues in the electrical industry with the updating of the book; in particular Mr. J.-H. Etula, Espoon Sähkö Oy, Mr. A. Mäkinen, Dr. K. Kauhaniemi and many others from Tampere University of Technology; Mr. H. Salminen of the Association of Finnish Electric Utilities; Professor J. Partanen, now at Lappeenranta University of Technology, and Mr. R. Hartwright of Midlands Electricity plc.

E. Lakervi                                                      E. J. Holmes
Kangasala, Finland                                   Stourbridge, England

*Chapter 1*

# The supply system

## 1.1 Generation, transmission and distribution

The primary aim of the electricity supply system is to meet the customer's demands for energy. Power *generation* is carried out wherever it gives the most overall economic selling cost. The *transmission* system is used to transfer large amounts of energy from the main generation areas to major load centres. *Distribution* systems carry the energy to the furthest customer, utilising the most appropriate voltage level. Thus an electricity supply system contains three different functions. Often individual supply organisations cover only one of these functions within a particular area or region.

Electricity is produced from a number of energy sources. Plant with a high capital cost, such as hydro and nuclear stations, is economic only if operated for most of the year at maximum output. On the other hand, plant which has a relatively low capital cost but high operating costs, such as gas turbines, may be more economic for short periods of operation to meet peak loads. Fossil-fuelled steam power stations, together with nuclear power plants and hydro stations, provide the majority of electrical energy. This is illustrated in Table 1.1, covering energy consumption in 1992 in some European countries, and in Figure 1.1 which shows the amounts of energy produced by these three sources in the European Community countries, the USA, the former USSR and Japan in 1991. Combined heat and power stations are favoured in some countries. In a combined heat and power station the energy normally lost in the condensation process is utilised by heat consumers. Thus the overall efficiency of the power station can be markedly increased. Diesel engines, wind and wave machines, and solar cells provide only a marginal amount of electricity on a world-wide basis, although their output may be significant locally. This kind of dispersed generation is rapidly increasing in some countries, however.

In any one country it is essential that resources be channelled into constructing generating plant which results in the lowest possible energy costs, taking into account the capital investment, and operating and maintenance

*Table 1.1   Sources of electrical energy (TWh) in European countries (1992)*

| Country | Net generated | Steam | Nuclear | Hydro | Net import | Total |
|---|---|---|---|---|---|---|
| Austria | 43·4 | 10·4 | — | 33·0 | 0·7 | 44·1 |
| Belgium | 68·4 | 26·1 | 41·1 | 1·2 | — | 68·4 |
| Denmark | 28·2 | 27·4 | — | 0·8 | 5·7 | 33·9 |
| Finland | 55·0 | 21·8 | 18·2 | 15·0 | 8·2 | 63·2 |
| France | 434·7 | 45·7 | 321·3 | 67·7 | −53·2 | 381·5 |
| Germany | 393·6 | 227·3 | 149·2 | 17·1 | −3·7 | 389·9 |
| Great Britain | 295·7 | 222·7 | 68·4 | 4·6 | 16·7 | 312·4 |
| Greece | 31·2 | 28·8 | — | 2·4 | 0·6 | 31·8 |
| Ireland | 14·6 | 13·9 | — | 0·7 | — | 14·6 |
| Italy | 214·2 | 169·1 | — | 45·1 | 5·2 | 249·4 |
| Luxembourg | 1·2 | 0·6 | — | 0·6 | 3·9 | 5·1 |
| Netherlands | 60·3 | 56·7 | 3·6 | — | 8·6 | 68·9 |
| Portugal | 26·3 | 21·3 | — | 5·0 | 1·3 | 27·6 |
| Spain | 140·5 | 67·8 | 53·4 | 19·3 | 0·6 | 141·1 |
| Sweden | 141·0 | 7·5 | 60·8 | 72·7 | −2·1 | 138·9 |
| Switzerland | 57·3 | 1·5 | 22·1 | 33·7 | −3·7 | 53·6 |
| Former Yugoslavia | 62·8 | 37·5 | 3·8 | 21·5 | −1·7 | 61·1 |

costs, subject only to proper environmental safeguards. The facility to operate power stations utilising different types of generation plant enables the load pattern to be matched better and gives more flexibility to achieve the lowest overall cost of electrical-energy production in the uncertainties of the future. For example, in Finland nuclear and combined heat and power stations each produce approximately one-third, hydro stations one-fifth and conventional condensate power stations only one-tenth of the total energy requirements.

To obtain maximum overall efficiency, combined heat and power stations should be located within a town, where the houses can be heated from the station hot water output, or within a factory requiring steam. Large rivers, where the drop in water level makes it possible to construct and operate larger hydro power stations, are often located in remote locations at considerable distances from the main areas of electricity demand. Similar considerations apply to geothermal stations and, for safety reasons, to nuclear stations. It is essential to have an adequate electrical system to transport electrical power from these large power stations to the main load centres. The demand can then be met by running the precise amount of power plant, wherever situated, operating at the most efficient loading to give the minimum overall cost under all credible system conditions. A strong transmission system is often justified economically because the difference in cost between different methods of generation is usually higher than any additional transmission cost.

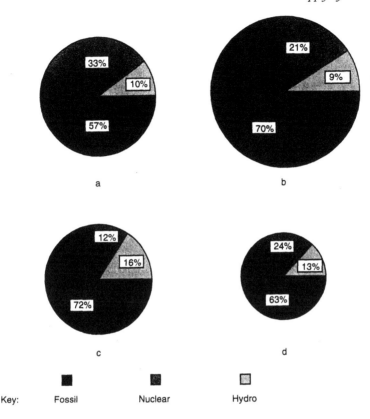

*Figure 1.1   Generation of electricity by energy source in 1991*

    *a*  EC countries, 1846 TWh
    *b*  USA, 3025 TWh
    *c*  Former USSR, 1680 TWh
    *d*  Japan, 838 TWh

Where the transport of very large amounts of power over large distances is involved, a very high-voltage system, sometimes termed major or primary transmission, is required. Such systems operate in the 300 kV plus range, typical values being 400, 500 and 765 kV. When transmission systems operating at lower voltage levels (110 or 132 kV) become overloaded, a higher-voltage system can be added as an overlay. An example of this is the 275/400 kV supergrid system established in Great Britain in the 1950s when the existing 132 kV system became incapable of coping with the large power transfers from the north to the south of England.

The following definitions for various voltage levels are used in the book. Low voltage (LV) is below 1 kV, while medium voltage (MV) covers voltage levels generally lying between 1 and 36 kV, used particularly in distribution systems, with high voltage (HV) being used for systems over 36 kV. In any electrical system the concept of LV, MV and HV has a relative meaning and does not necessarily comply with national safety or other regulations. The term EHV is used in this book to signify voltage levels above 300 kV.

EHV interconnections have been constructed between regions and countries in order to achieve the most economic use of generation and transmission plant overall, and to cater for temporary shortages of plant in any one region. In Europe the largest of such arrangements operates under the auspices of the Union for the Co-ordination of the Production and Transmission of Electrical Energy (UCPTE). This is shown schematically in Figure 1.4, where the width of the lines indicates the relative energy transfers between the various countries and the size of the circle indicates roughly the amount of electrical energy consumption in 1992. Owing to the political situation in former Yugoslavia in the 1990s, parts of the Yugoslavian network and Greece were not synchronised to the UCPTE network. Power exchanges with Nordic countries are co-ordinated via the NORDEL organisation. As will be seen from Figure 1.4, further interconnection facilities exist with Russia, the Czech Republic, Hungary and Romania. Owing to the former political situation in Europe and the many technical differences between the western and eastern power systems, DC links have been used to interconnect these systems. A DC link is in operation on the interconnection between NORDEL and UCPTE and HV DC links also interconnect Great Britain and France. Similar EHV interconnectors link various regions in the USA and Canada to pool generation capacity. The political changes in eastern Europe and the increasing activities in the whole

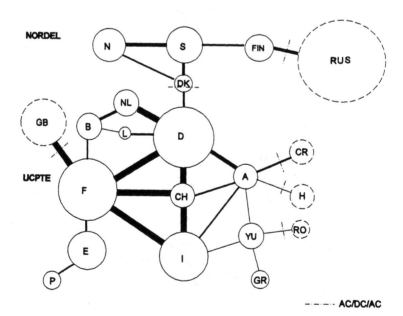

*Figure 1.4   The Western European UCPTE and NORDEL interconnection ties*

world towards new and larger interconnected systems have also raised the question of the need for, and technical possibilities of, connecting the western and eastern electric power systems.

Depending on the number, and size, of conductors per phase and the distance involved, a single 400 kV transmission circuit could carry the output of a 2000 MW power station. If this output were lost for just one hour, the resultant costs of any disruption of supply could run into millions of pounds, thus justifying a high level of security – for example installing parallel circuits to improve reliability and stability. It is also possible that an industrial load of, say, 20 MW supplied from the HV system would need to be secured against a single fault if this would detrimentally affect the customer's production or equipment. EHV and HV systems are therefore planned so that single-circuit faults rarely result in loss of supply to any customer. This assumes effective protection systems and adequate alternative back-up connections.

The HV networks may be operated as interconnected systems or discrete groups. When operating as interconnected systems these networks may then be used to provide back-up capacity to the higher-voltage networks.

Below the transmission system there can be two or three distribution voltage levels to cater for the variety and demands of customers requiring electricity supplies. In general, the MV and LV networks are operated as radial systems. The theoretical and technical aspects of planning and operating these systems, as well as the supporting HV systems, form the major part of this book.

## 1.2 Supply organisations

Since the activities associated with the three functions of electricity supply, i.e. generation, transmission and distribution, are so widely different, this provides a basis for splitting these activities between different supply organisations. Usually the main generation and transmission activities are carried out by the same undertaking, although combined heat and power stations are often owned by industrial companies or municipal authorities.

In the 1990s there has been a strengthening trend towards breaking up the vertical integration in the electric power industry by separating the generation, transmission and distribution of electricity into separate business areas. This trend includes the increasing demand towards opening up transmission and distribution networks to producers and customers and the appearance of independent power producers. Until now, many electricity undertakings have been considered to be 'public services' rather than commercial businesses. The ultimate goal in the ongoing 'deregulation' process is to consider electricity as a product that can be sold in a free market without any restrictions from official authorities.

In the European Community, as part of the deregulation process, the first legislative action with the objectives of increased integration, removal of barriers to trade, improved security of supply, reduced costs and enhanced competitiveness was the adoption in 1990 of two Directives. The Transit Directive gives the right to electricity utilities to import or export energy across a third country, and the Price Transparency Directive aims to ensure that the energy market is not distorted by hidden subsidies.

In some countries electricity distribution supply organisations form part of a nationalised energy authority, controlled via a government department. Distribution systems may also be owned and operated by the local town administration or council. Additionally, distribution systems may also be owned by private companies and others have a mixture of public and private ownership. A distribution supply company could have a monopoly of supplying electricity within a given area which, in the extreme, may encompass a whole country, while at the other end of the scale it may cover only a small town. The strengthening trend towards separating electricity generation from transmission and distribution and the demands of opening a free market for electricity, however, will lead to the reorganisation of power utilities.

In Europe deregulation of the electricity supply industry is most advanced in the United Kingdom and in some NORDEL countries. In Norway, electricity transmission has been separated from generation and a free market has been opened for electricity, while a similar development is taking place in Finland and Sweden in 1995 and 1996, respectively. In the power distribution sector, deregulation means that customers are allowed to buy their energy from any generating or selling organisation. Thus a traditional distribution company or utility must now divide its activities between the energy and network business areas, with the energy business deregulated and operating under normal commercial competition, while the network activities will remain more or less monopolistic. In addition, it is foreseen that new companies will be set up in the energy business area. In the network business area regulations are required to ensure that the power transfer through any distribution network is reasonably priced.

In Finland one state-owned company accounts for nearly one-half of the country's generation, with private industry, power companies and municipalities providing the rest of the generation. A separate state-owned company is responsible for the main part of the 400 kV transmission system, while the 220 kV and 110 kV systems are owned by several organisations. Electricity distribution is provided by some 100 private or municipal companies, with rural parish communities often being the main share owners in the private companies supplying their area.

Of the EC countries only Great Britain has reorganised its electricity supply industry by transferring it to private ownership; it now consists of companies formed from the previously state owned organisations plus a number of other smaller single station generating companies. In England and Wales, with the exception of nuclear energy production which is still in public ownership, electricity is mainly generated by two private companies. One organisation is responsible for electricity transmission, and 12 regional electricity companies have a responsibility for distribution within their authorised areas. In Scotland the two public organisations are now privatised, operating as vertically integrated companies covering generation, transmission and distribution. Here, also, nuclear generation remains in state ownership.

In the Federal Republic of Germany some 1000 electricity supply companies are involved in electricity generation, transmission and distribution for public supplies, with joint-stock companies predominating. The nine interconnected power companies cover 80% of the public supply demand. Some industrial companies generate electrical energy for their own demand, with any surplus energy being made available to the public network. 90% of the transmission line network (the EHV network) is owned by the nine interconnected power companies, with 55 regional companies being responsible for regional transmission and distribution.

At the time of writing, Italy was moving towards privatisation of its generation, transmission and distribution systems. Elsewhere Argentina, Colombia, New Zealand and several eastern European countries have already taken significant steps to open their markets.

In France generation, transmission and distribution of electricity are still covered by the nationalised organisation Electricité de France (EdF). Some independent generation takes place outside EdF, for example associated with the French railways and coal mines, and a few distribution networks operated by certain public and co-operative organisations have also been left out of the nationalised industry.

Within the USA there is a wide spread in the size and type of electricity utilities. There are more than 3000 individual undertakings engaged in the generation, transmission and distribution of electricity, or combinations of these. Just under 200 private-investor-owned utilities form the largest segment of the electricity supply industry, serving about 75% of the retail customers. There are also some 2700 municipal, state or county utilities. Approximately 900 co-operatives are owned by groups of people who have organised joint ventures for the purpose of supplying energy to specified rural areas. The US Federal Government operates seven big utilities, two examples being the Tennessee Valley Authority (TVA) and the Bonneville Power Authority (BPA).

## 1.3 The distribution system

The function of an electricity distribution system is to deliver electrical energy from the transmission substations or small generating stations to each customer, transforming to a suitable voltage where necessary.

Figure 1.5 illustrates the interrelation of the various networks. The HV networks are supplied from EHV/HV substations which themselves are supplied by inter-regional EHV lines. HV/MV transforming substations situated around each HV network supply individual MV networks. The HV and MV networks provide supplies direct to large customers, but the vast majority of customers are connected at LV and supplied via MV/LV distribution substations and their

In rural areas overhead MV lines are in general use. While some earlier installations using copper conductors remain, it is now common practice to use 25–100 mm² steel-cored-aluminium conductors and wooden poles. Aluminium-alloy conductors now account for about 10% of the total length of conductor installed each year. Concrete and steel poles are used in regions where wood is not readily available or climatically suitable. Insulated conductors are more reliable and also environmentally more acceptable and are tending to be used increasingly. Of necessity underground cables are being installed more frequently in urban areas, but they are only used in rural areas for environmental reasons, e.g. to avoid a wirescape effect close to a substation.

Pole-mounted transformers, with ratings from 5 to 315 kVA, form the rural distribution substations. The smaller sizes are sometimes single-phase units. In urban areas supplied by MV underground cables the substation can be at ground level, in brick, steel or concrete enclosures, or placed in the basements of office or housing blocks with transformer sizes varying from about 200 to 2000 kVA.

Low-voltage distribution in rural areas is by means of overhead lines with bare or insulated conductors. The use of aerial bunched conductors is becoming dominant, particularly in villages, small towns and forested areas. The length of new LV lines is usually limited to around 500 m or even less, depending on such factors as the voltage, the number of phases used and the load. In town centres underground cables predominate. In urban areas back-up supplies are often available from neighbouring distribution substations.

In those countries where the provision of electricity supply has just started, it is essential that the choice of system design philosophy be appropriate to the local social and economic conditions. In any 'green field' situation the design engineer has the opportunity of introducing the most suitable technology available for equipment, automation, telecontrol and microprocessor applications, which can have considerable advantages in sparsely populated areas with long distances between substations, and often over difficult terrain. Changing from one practice to another, even if technically better, is seldom economic, at least in the short term. It is particularly important that a suitable supply organisation be set up, and that a system design philosophy which is most suited to an individual country's own conditions be developed. Those factors which have an influence on the choice of voltage levels, system earthing, the use of three-phase or single-phase supplies, and operational and protection arrangements are discussed in subsequent chapters.

## 1.4 Supply requirements

With the increasing dependence on electricity supplies the necessity to achieve an acceptable level of *reliability*, *quality* and *safety* at an economic price becomes even more important to customers. The price that a customer has to pay for electrical power is dictated by the costs of the associated generation, transmission and distribution systems. Particularly for those customers whose energy demand is high, the tariff level is a vital component of their overall costs and may well influence whether electricity or some other form of energy is used.

Virtually all customers consider the reliability of the electricity supply to be of major importance, affecting as it does the working, domestic and social aspects of their life. Whilst the customer would like to have total reliability and never to lose supply even for a brief period, supply authorities are aware that technically and financially this would be an impossible target. The customer is concerned about any financial loss or inconvenience he or she may suffer as a consequence of a supply interruption. The effects of loss of supply vary with the customer group. A number of studies carried out indicate the wide range of values that customers place on loss of supply (outage). Figure 1.7 gives one example of such studies.

Figure 1.8 indicates the costs incurred by customers as a result of loss of supply. With 100% availability, i.e. total reliability, these costs would be zero. An electricity utility incurs additional expenditure as the reliability is improved. Eventually these costs increase dramatically for each additional percentage-point improvement, so that to try to achieve 100% reliability would be economically impracticable. On the basis that the total minimum cost to the

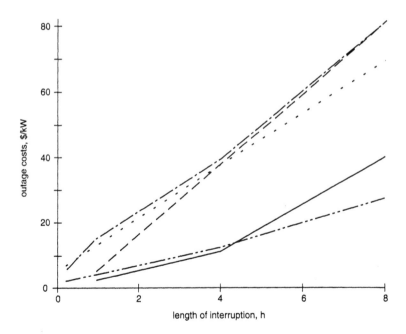

*Figure 1.7 Estimated values of outage costs for different customer groups (courtesy VTT Energy)*

| ——— residential | - - - - industry | ..—--— public sector |
| — — agriculture | --—--— commercial | |

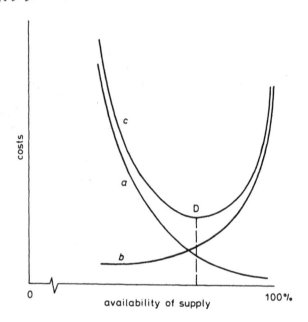

*Figure 1.8   Balance between utility and customers' costs*

    *a* Cost incurred by customers as a result of loss of supply
    *b* Cost incurred by utility in providing the availability of supply
    *c* Total cost = *a* + *b*

community should determine the level of availability, this would be at point D
on curve c.

As the costs of unreliability tend to be related to the electrical demand not
supplied, and to the length of the interruption, recommendations concerning
standards of security should perhaps best be related to these factors. The
combined load of a group of customers or substations affected by a fault on the
distribution system could be used as an indicator against which to set down
various levels of system security. Without specific recommendations the
tendency would be that 'good supplies' get better and 'poor supplies' get worse.

Network reliability can be considered as a cost factor, similar to investment or
losses. If estimated outage times and loss of demands were calculated for each
customer, then, using the costs in Figure 1.7 or some equivalent cost factor, it
would be possible to obtain a measure of reliability. In practice this type of
reliability calculation can be carried out using computers; these topics will be
covered in Chapter 4; and Sections 9.5 and 14.6.

Another factor affecting the quality of supply is the actual value of the supply
voltage, which needs to be kept within a given range for the correct operation of
customers' appliances. With extreme excursions outside the accepted voltage
range, it is possible that some appliances could be badly damaged. Excessively
high voltage is usually due to failures in the voltage control equipment, or the
result of system faults producing overvoltages. Too low a voltage generally

results from excessive voltage drops on the distribution networks. In addition to variations in the actual value of voltage being supplied, the voltage waveform may vary from a pure sine wave and thus cause improper operation of utility equipment or customers' appliances. Some of the causes of waveform distortion and other phenomena in respect of user voltages are discussed in more detail in Chapters 12 and 13.

The third requirement highlighted at the head of this Section is the safety of the electricity supply. The power-distribution systems themselves may cause danger to people, animals and property in a number of ways unless suitable precautions are adopted. Customers' appliances must be constructed and operated in such a way that they do not lead to accidents involving, for example, electric shock, or fires caused by overcurrents or faults. Safety can be improved in various ways, e.g. by ensuring that adequate clearances are maintained between conductors and earth (ground), using an appropriate method of earthing the power network, and by providing suitable reliable protection on all circuits and electrical equipment. Some of the safety requirements may be dictated by public authorities but the codes of practice adopted by utilities, including those for design, generally ensure that actual safety standards are better than the minimum required by law. Nonetheless, absolute safety cannot be achieved, so that efforts for improved safety must all be co-ordinated with other supply requirements.

Problems of limited finance and other resources can lead to strategies being adopted to extend supplies as fast as possible whilst accepting some tolerances on the quality and standards of supply. The standards can then be improved as the electricity system develops and additional finance becomes available.

## 1.5 Network configurations

In designing distribution networks supplies can be provided to different areas of the system in a variety of ways, depending on the load density and system voltage level. Various arrangements of mesh and interconnected networks are illustrated in Figure 1.9 where individual substations are represented by circles. The interconnecting circuits of the mesh arrangement can provide increased security of supplies to individual substations, and this arrangement is therefore frequently used in HV systems. While this arrangement requires more substation equipment overall, for example switchgear and electrical connections, it is usually more efficient in terms of total circuit costs. The mesh arrangement is easier to extend and has a higher utilisation of circuits when fully developed than a ring system, although this can result in higher network losses. Rings and mesh systems may be operated split with normally open points, or closed to improve security, although the latter require more circuit breakers and more sophisticated protection.

By interconnecting and operating a number of infeed substations in parallel as shown in Figure 1.9*b* it is possible to reduce the total transformer capacity into

*Telecontrol systems* enable real-time information to be obtained from the supply system and permit remote-control operation of various switching equipments. Such systems thus actively assist in improving fault clearance times and the overall security of supply. Microprocessor-based modular telecontrol systems make it possible to monitor and control individual remote items of equipment to achieve a better operational standard.

The information collected via the telecontrol system can easily be processed and stored in data banks and then later used as the basis for network design studies. The telecommunication network is an essential part of any telecontrol system, and public and utility-owned telephone networks, radio links, power-line carrier systems as well as optical-fibre arrays now offer alternative data-transmission paths. Applications of *distribution automation* are expanding rapidly. Typical conventional examples of these at substations are alarm centre, transformer voltage-control units, tuning systems of Petersen coils, interlocking systems and sequence control of switchgear. These can also be integrated into the telecontrol system. Network information systems, which traditionally have been used for computer-aided design of distribution networks, are now capable of introducing useful support to automation applications and are being used more and more, for example in providing facilities for automatic fault location and restoration. Load management systems have been developed to reduce the peak loads on distribution network equipment, although the major benefit of such systems is in reducing the peaks of generation output. Various forms of signalling are in use to switch customers' loads on and off as necessary, including coded ripple signals injected into the distribution network and radio transmitted signals to tuned receivers in customers' premises.

By using or introducing some or all of these auxiliary systems the design engineer has the facility for improving network utilisation, reducing customer supply interruptions, and ultimately reducing capital and revenue expenditure.

*Chapter 2*
# Planning distribution networks

## 2.1 Overall philosophy

The planning and design of electricity distribution networks can be divided into three areas. Strategic or *long-term planning* deals with future major investments and the main network configurations. *Network planning* or design covers individual investments in the near future while *construction design* includes the structural design of each network component taking account of the various materials available. The object of this chapter is to introduce those factors which should be taken into account when designing electricity distribution systems. The emphasis will be on general planning guidelines. Later chapters are devoted to more detailed considerations of technical and economic aspects, and specific engineering topics.

Long-term planning of the distribution system is an essential part of the planning activities of an electricity supply utility. Its main purpose is to determine the optimum network arrangements, what investments would be required, and the timing of these to obtain maximum benefits. At each stage the appropriate regulations covering such matters as quality of supply, safety and amenity should always be met while keeping the total costs over the life of the system as low as possible. To achieve this all the cost components – not just capital investments and their timing but also continuing annual costs such as system losses and maintenance expenditure – must be taken into account.

In industrialised countries the existing supply system covers virtually all the inhabited areas. In these areas the existing network usually provides a good starting point for planning future system arrangements. The need for further investment is usually to cater for load growth, or to replace ageing assets on the network. Uprating existing overhead lines and other plant, where technically feasible, can often be more economic than installing new circuits or equipment when the costs of obtaining new line routes or new substation sites are taken into account. However, the longer-term cost of developing the network must also be examined. The above objectives may be modified as time passes. In addition, it will be necessary to consider the effect of future changes in such factors as the growth of load in different areas, variations in the relative levels of the costs of materials and energy, as well as the increasing use of technical innovations.

From these planning studies it is possible to draw up lists of expected major annual investments and also to allocate sums of money to cover future undefined technical schemes. The total investment requirements must then be compared with an associated financial plan which takes account of assumptions made concerning the bulk purchase prices of electricity, load growth, system losses, existing loans, new borrowing requirements and any changes in salary costs, safety requirements etc. Over the period of the financial plan a balance should be achieved so that both the technical and financial objectives are reached within a stable and acceptable energy-sales tariff level.

The planning activity often includes not only determining major future network requirements, e.g. additional primary substations or the major revision of a telecontrol scheme, but also producing planning guidelines. These are usually intended to cover the smaller and more common types of investment for which it would not be economic to carry out individual technical and economic appraisals in detail; e.g. the replacement of distribution transformers or overhead-line conductors, or the design of housing-estate supplies. In this way common policies can be more easily implemented. Economic aspects often play an important role in determining such guidelines since, in addition to any technical factors, the costs of alternative policies also have to be taken into account when drawing up these guidelines. The written guidelines used by utilities are increasingly being transferred to network information systems, which are discussed in Chapter 14.2.

## 2.2 Planning objectives

A large variety of regulations, guidelines and recommendations are produced to ensure that electricity supply systems can be properly operated for the benefit of society. In most countries regulations concerning safety, the supply arrangements and the proper use of electricity form part of governmental legislation and in Europe, especially in the EC countries, efforts are being made to bring regulations together in a unified manner. Rules or recommendations for including economic optimisation in design procedures, for example, or for obtaining a suitable level of income to cover future investments are best determined by the utility management or owners. Although supply utilities have some degree of monopoly within their distribution areas, their economy should be balanced in such a way that the unit tariff level and profit margin are satisfactory to customers and owners, respectively. Whether the utilities are owned privately, or by municipal or state authorities, or by any combination of

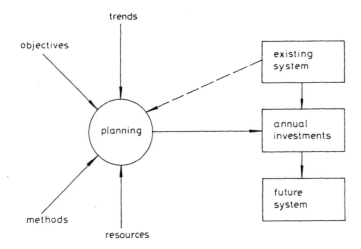

*Figure 2.1    Planning as part of the distribution-system development process*

these, does not in principle affect the situation, although municipal and state authorities are often more subject to political influence.

Planning objectives concerning the quality of supply, tariff price levels and stable employment are in common use in industrialised countries. There are still large variations from one country to another regarding the level at which regulations are formulated, and whether these are mandatory or obligatory, and also in the degree of detail of the regulations. This applies whether or not the regulations are laid down by local or state authorities or by the utilities themselves.

The detailed requirements of individual technical regulations may have a significant effect on such matters as the quality of supply, safety and the costs of providing electricity supplies to customers. For example, regulations setting out the minimum acceptable ratio of the lowest single-phase short-circuit current to the rating of a protection fuse on the low-voltage network will tend to increase costs, especially in rural areas. On the other hand this requirement improves safety to the system and customers, and reduces voltage fluctuations. This therefore encourages more rational investment than would, for example, occur if an extreme value were chosen for the minimum voltage to be received by any customer.

The number of national regulations and recommendations plus the internal design policies and engineering recommendations of a utility can be large. Considerable effort is required to ensure that all these documents are regularly updated to take account of improved techniques or changes in policy. It is nevertheless essential for a utility to ensure that everyone involved in the design of its supply networks has the most up-to-date information and tools and uses them correctly.

than to have to make major revisions to distribution networks or their associated auxiliary systems at some later date.

If major variations can be foreseen in the cost relationships between various electrical components compared to their capability and energy losses, these should also be considered at the investment appraisal stage. One example is the introduction of low-loss transformers where it may not be possible to justify replacement solely on savings in losses, although the saving in losses may be a factor in determining any replacement of obsolescent or deteriorating transformers. The use of XLPE cable, with a higher conductor operating temperature, permits the use of a smaller cross-sectional area conductor than with the older cables. However, this reduced cross-sectional area results in an increased resistance, and therefore higher $I^2R$ losses. It is therefore necessary to take account of the capitalised losses as well as the initial investment cost when comparing the economics of XLPE cables against those of existing cables when considering network reinforcement.

## 2.6 Production of long-term plans

At HV the technical studies can be complex, especially if generation is involved. The design and construction period often includes complicated negotiations to obtain routes for overhead lines and underground cables. Consequently the overall lead time for individual HV schemes, from the planning stage to completion, is so long that most of the work load and expenditure up to the end of the review period, say eight years ahead, can be reasonably accurately estimated. For MV networks it is usually possible in the earlier years of the review period to divide the future work programme into those required for specific projects and those to cater for general load growth. From this an assessment can be made of the engineering and financial resources required for at least half the review period. Thereafter it may be necessary to make provision for future schemes as yet unknown, in addition to any known large projects.

All these considerations will involve some combination of network reinforcement and extension by circuits and units of large capacity, and the addition of new substations and circuits. If the timing, and the size, of the investment in HV and MV projects are determined from cost/benefit analyses of the various options available for each project, the most cost-effective schemes can be included with some confidence in both the engineering and financial long-term plans. This is an important aspect when such schemes take up a considerable proportion of a utility's financial and manpower resources.

On the other hand, because of the short lead time usually given to a utility by customers and housing and industrial developers requiring new or additional supplies, low-voltage networks cannot be investigated in such detail over such a long time period. Nonetheless, studies based on estimated loads for the next few years ahead will indicate the number of feeders likely to require reinforcement, generally due to excessive voltage drop, and those cases where there could be insufficient short-circuit current to operate protective devices satisfactorily. Such studies will indicate particular requirements for some years ahead. For the later years of the programme some general provision will need to be made for the LV systems based on past experience, perhaps related to overall load growth, for example.

Allowance must also be made in such plans for the replacement of equipment at all voltage levels, together with offices, shops and vehicles, and miscellaneous equipment such as tools and computers. For equipment which has a long useful life, such as cables, allowance must be made to write off any equipment which becomes surplus to system requirements, e.g. owing to network reorganisation or loss of load, before such items reach the end of their useful technical life and require replacing. This may occur in a system having two voltage levels, where rationalisation results in one voltage level being removed, with redundant cables left in the ground which cannot be re-used in the revised network configuration designed for expected future developments. In some circumstances there may be an economic case for purchasing equipment at a cheaper capital cost whilst accepting that its useful life may be shorter. However, the economic assessment should then take account of any increased maintenance costs and losses, together with earlier replacement when the shorter technical lifespan is reached.

In addition to the estimated cost of reinforcing and maintaining a utility's electrical system, it is necessary to include other future costs such as the provision of an adequate transport fleet, and possible activities associated with combined heat and power systems. The production of long-term plans thus necessitates a financial plan requiring co-operation between all the main sections of a utility. The annual updating of these plans and the planning and design policy documents provide the necessary support system for the detailed designing of a utility's electricity supply system.

## 2.7 Bibliography

ANDERSSON, I., BUBENKO, J. A., and LIVEUS, L.: 'Future aspects on distribution network planning – The Swedish experiences'. 7th International Conference on Electricity Distribution, CIRED 1983, AIM, Liege, Paper *a* 14

BEG, D.: 'Electric sector planning: a comparative experience in Asia and Pacific region', *Power Eng. J.*, 1987, **1**, pp. 133–142

BERRIE, T. W.: 'Power-system planning under uncertainty – a new approach', *Power Eng. J.*, 1987, **1**, pp. 92–100

BERRIE, T. W.: 'Electricity economics and planning' (Peter Peregrinus Ltd. on behalf of the IEE, London, 1992)

'Data sheet: World electricity supplies', *Elect. Rev.*, 1986, **219**, (2)

DAVENPORT, F. W. T.: 'Electricity supply – distribution at the crossroads', *Power Eng. J.*, 1988, **2**, pp. 7–15

DODDS, G. I., and BEATTIE, W. C.: 'Effects of abnormal operating conditions on electrical utility planning', *IEE Proc. C*, 1991, **132**, (2), pp. 113–120

FREDRIKSEN, O.: 'The design of distribution networks for rural areas'. IEE Conf. Publ. 151, International Conference on Electricity Distribution – CIRED 1977, pp. 160–163

GÖNEN, T.: 'Electric power distribution system engineering' (McGraw-Hill, 1986)

HAMMERSLEY, H.: 'Design and development of public distribution supply systems', *IEE Proc. C*, 1986, **133**, pp. 370–372

HOLLAND, R.: 'Appropriate technology – rural electrification in developing countries', *IEE Rev.*, 1989, **35**, (7), pp. 251–254

MARSHALL, A. C., BOFFEY, T. B., GREEN, J. R., and HAGUE, H.: 'Optimal design of electricity distribution networks', *IEE Proc. C*, 1991, **138**, pp. 69–77

McCALL, L. V.: 'North American versus European Power Systems'. CIGRE Study Committee 36, July 1987

MENON, S. G. and RAO, B. B. V. R.: 'Planning of distribution systems in developing countries', *IEE Proc. C*, 1986, **133**, pp. 384–388

PANSINI, A. J.: 'Electrical distribution engineering' (McGraw–Hill, 1983)

TELLIER, R., and GALLET, G.: 'French distribution systems: basic features and practices', *IEE Proc. C*, 1986, **133**, pp. 377–383

*Chapter 3*
# Technical considerations

## 3.1 General

Good system planning and design requires a sound knowledge of the existing electrical system to provide a firm base on which to assess projects for future network development. Technical factors such as those aspects which can influence future loads need to be considered. However, it is essential that any project under consideration by a planning engineer must not only be examined from the point of view of various technical aspects; a comprehensive economic assessment of each project must also be carried out at the same time, to ensure that any proposed development is technically sound and cost effective.

This chapter covers various technical aspects which should be considered for both normal and abnormal operating conditions. The planning engineer has to consider the effect of the loss of any item of equipment on the supplies to customers and on the quality of supply, e.g. voltage fluctuations, and the amount of time a customer may be off supply, as well as the safety of the public and the utility staff. It is also necessary to take account of the effect of transient and permanent system faults on both utility- and customer-owned equipment.

## 3.2 Modelling network components

### 3.2.1 Components

When carrying out technical calculations it is necessary to make use of 'equivalent circuits' for various components, and then combine these circuits in order to represent the interconnection of the components in the actual electrical network. The relatively short lengths of MV and LV distribution circuits enable simple modelling techniques to be used for lines. The radial network configurations usually used make it possible to simplify the network model, and large matrix models are seldom necessary. It is usually sufficient to represent a distribution circuit by a series impedance and ignore its capacitance, except

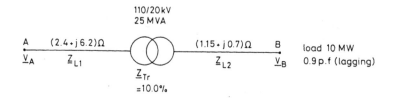

*Figure 3.1   Example network*

where the load at B is fed from source A via a 110 kV line, a 110/20 kV transformer and a 20 kV feeder. The load voltage is assumed to be 20 kV. It is required that the source voltage at point A be determined.

In the following calculations, transformed impedances and voltages to the 20 kV side have been marked by the symbol $'$.

$$Z'_{L1} = \frac{(20)^2}{(110)^2}(2{\cdot}4 + j6{\cdot}2)\Omega = (0{\cdot}079 + j0{\cdot}205)\,\Omega$$

$$Z_{Tr} = j0{\cdot}10\frac{(20)^2}{25}\Omega = j1{\cdot}6\,\Omega$$

$$Z'_{AB} = Z'_{L1} + Z_{Tr} + Z_{L2} = (1{\cdot}23 + j2{\cdot}51)\,\Omega = 2{\cdot}80\underline{/63{\cdot}89^\circ}\,\Omega$$

$$P = \sqrt{3}VI\cos\phi$$

$$I = P/\{\sqrt{3}V\cos\phi\} = 10 \times 10^6/\{\sqrt{3} \times 20 \times 10^3 \times 0{\cdot}9\}\,A$$

$$= 320{\cdot}8\,A$$

$$I = 320{\cdot}8\underline{/-25{\cdot}84^\circ}\,A$$

$$V'_A = V_B + \Delta V \simeq V_B + \sqrt{3}I\,(R'_{AB}\cos\phi + X'_{AB}\sin\phi)$$

$$= 20\,000 + \sqrt{3} \times 320{\cdot}8(1{\cdot}25 \times 0{\cdot}9 + 2{\cdot}51 \times 0{\cdot}436)\,V$$

$$= 20\,000 + 1220 = 21\,220\,V = 21{\cdot}220\,kV$$

$$V_A = (V_{1n}/V_{2n})V'_A = (110\,000/20\,000) \times 21{\cdot}220 = 116{\cdot}7\,kV$$

### 3.2.3 Per-unit values

In more complex networks, e.g. those including more than two voltage steps, the above impedance-transforming method can prove tedious. The use of per-unit values, e.g. ratios of actual values to certain base values, can be usefully employed in overcoming problems of transforming impedances across different voltage levels. The benefits of using per-unit values are:

- results for different systems are comparable, e.g. voltage drop $v_d$ and power losses $p_l$
- transformer impedances are identical for both sides
- $\sqrt{3}$ factors are not needed in 3-phase calculations.

Usually the base value for the apparent power $S$ is fixed to a typical transformer rating, and that for $V$ to the nominal voltage. The base voltage $V_b$ for other voltage steps is calculated by using the transformer ratios, while $S_b$ is the same throughout the network being studied. $I_b$, $Z_b$ and $Y_b$ can be calculated from $S_b$ and $V_b$.

The example in Section 3.2.2 is now used again in order to illustrate the use of the per-unit method. For calculating per-unit values the base power will be taken as the rated power of the transformer, i.e. $S_b = 25$ MVA, and is common to both voltage levels.

Calculating the base values of current for a 3-phase system:

$$\text{at } 110 \text{ kV}, \quad I_{bA} = S_b/\sqrt{3}V_{bA}$$

$$= (25 \times 10^6)/\sqrt{3} \times 110\,000 = 131\cdot22 \text{ A}$$

$$\text{and at } 20 \text{ kV}, \quad I_{bB} = S_b/\sqrt{3}V_{bB}$$

$$= 721\cdot7 \text{ A}$$

and similarly the base values of impedance are:

$$\text{at } 110 \text{ kV}, \quad Z_{bA} = V_{bA}^2/S_b$$

$$= (110\,000)^2/(25 \times 10^6) = 484 \,\Omega$$

$$\text{and at } 20 \text{ kV}, \quad Z_{bB} = V_{bB}^2/S_b$$

$$= (20\,000)^2/(25 \times 10^6) = 16 \,\Omega$$

The line ohmic values are then converted to per-unit values:

$$z_{L1} = \frac{(2\cdot4 + j\,6\cdot2)}{484} = 0\cdot05 + j\,0\cdot013$$

$$z_{L2} = \frac{1\cdot15 + j\,0\cdot7)}{16} = 0\cdot072 + j\,0\cdot044$$

The transformer reactance, given as 10% on 25 MVA, is thus $0\cdot1$ p.u. on the chosen base of 25 MVA, so that the total per-unit impedance between A and B is $0\cdot077 + j\,0\cdot157$.

A load of 10 MVA at $0\cdot9$ power factor, at 20 kV, equates to current $I$, where

$$I = (10 \times 10^6)/(\sqrt{3} \times 20 \times 10^3 \times 0\cdot9)$$

$$= 320\cdot8\underline{/-25\cdot84°} \text{ A}$$

$$= 288\cdot7 - j\,139\cdot8 \text{ A}$$

$$= I_p + j\,I_q$$

Therefore the per-unit value of load current is

$$i = I/I_{bB} = (288\cdot7 - j\,139\cdot8)/721\cdot7 = 0\cdot400 - j\,0\cdot194$$

The per-unit voltage drop $v_d$ at busbar B is

$$v_d \simeq i(r \cos \phi + x \sin \phi)$$
$$= i_p r + i_q x$$
$$= (0 \cdot 400 \times 0 \cdot 077) + (0 \cdot 194 \times 0 \cdot 157)$$
$$= 0 \cdot 061$$

equivalent to

$$V_d = v_d V_b$$
$$= (0 \cdot 061 \times 20\,000)$$
$$= 1220 \text{ V}$$

Thus, referring to Figure 3.1, $V_A \simeq V_B + V_d$ and, relating this to the higher voltage level, is given by

$$V_A = (110\,000/20\,000) \times (20\,000 + 1220) = 116 \cdot 7 \text{ kV}$$

In this simple example the impedance-transformation method has proved to be less laborious than the per-unit method. Which method should be preferred depends on the system being studied. For distribution networks having only one MV step, impedance transformation is generally more useful, while in double MV systems and transmission networks the per-unit method is usually preferred.

Sometimes the impedances of different equipment may be stated on various power (MVA) bases. Having calculated the per-unit impedance on one base this can be converted to any other base using the following formula:

$$Z_{pu1}/Z_{pu2} = S_{b1}/S_{b2} \tag{3.3}$$

## 3.3 Power flows and losses

### 3.3.1 Power flows

For distribution networks AC load-flow studies are necessary to determine the capability of a network under all loading conditions and network configurations. This includes taking account of the loss of one or more circuits or items of equipment including the infeed power sources, whether from generation within the network or from transformer substations where the infeed power is obtained from a higher-voltage network. Most MV and LV networks are operated radially. As a consequence studies on such networks are relatively simple. On the other hand, the number of load points per network is higher and the information on the individual points is often limited with only the annual unit consumption figures at low voltage being known.

The power flow through each section of a network is influenced by the disposition and loading of each node point, and by system losses. Maximum-demand indicators installed at MV network infeeds provide the minimum amount of load data required for system analysis. More detailed loading information of the incoming supply and the outgoing feeders is economically available through the use of microprocessor units and telemetry.

Sometimes the numbers and ratings of customers' equipment appliances, and thus their maximum possible demand, are known. However, in order to carry out power-flow studies on MV and LV networks it is necessary to apply correction factors to individual loads. This is because summating the maximum values of all the loads will result in too high a value for the total current flows, and therefore the overall voltage drop, if the loads do not peak at the same time. It is therefore necessary to derate each individual load so that the summation of the individual loads equals the simultaneous maximum demand of the group of loads. This is achieved by applying a *coincidence factor*, which is defined as the ratio of the simultaneous maximum demand of a group of load points to the sum of the maximum demands of the individual loads. The inverse of the coincidence factor is termed *diversity factor*.

If kWh consumption information is available then empirical formulas or load-curve synthesis can be used to determine demands at network node points. The derivation of load data, and the use of computers for network studies, are discussed in Chapters 11 and 14.

### 3.3.2 Power losses

When the maximum current or real and reactive power flows have been determined, the series active and reactive power losses in a 3-phase circuit or any item of equipment, $P_l$ and $Q_l$ can be calculated from the following equations:

$$P_l = 3I^2 R_l \tag{3.4}$$

$$\text{or} \quad P_l = \left[\frac{P}{V}\right]^2 R_l + \left[\frac{Q}{V}\right]^2 R_l \tag{3.5}$$

$$\text{and} \quad Q_l = 3I^2 X_l$$

$$= \left[\frac{P}{V}\right]^2 X_l + \left[\frac{Q}{V}\right]^2 X_l \tag{3.6}$$

where $R_l$ and $X_l$ refer to the circuit series resistance and reactance as shown in Figure 3.2.

Given that the circuit shunt impedance is $(R_s + jX_s)$, as indicated in Figure 3.2, the shunt losses can be calculated using the shunt current $I_s$ instead of $I$:

$$I_s = \frac{V/\sqrt{3}}{\sqrt{R_s^2 + X_s^2}} \tag{3.7}$$

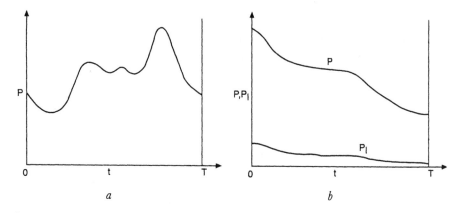

*Figure 3.3   Load and load-duration curves*

    *a* Load curve
    *b* Duration curves for power ($P$) and power losses ($P_l$)

The *load factor* $F$ is defined as the ratio of the average power divided by the maximum demand, and can be expressed as $W/P_{max}T$. The *loss load factor* is defined as the ratio of the average power loss divided by the losses at the time of peak load, expressed by $(W_l/P_{lmax}T)$. The load factor can be determined by integrating the duration curve for $P$, and the loss load factor by integrating the $P_l(t)$ curve. The quadratic relationship between $P_l$ and $P$ is shown in Figure 3.3*b*.

Where only the load factor $F$ is available, various formulas have been developed to obtain a quick approximation of loss load factor (LLF), generally based on the expression $\text{LLF} = aF + (1 - a)F^2$. Two examples are given below:

$$\text{Loss load factor} \simeq 0{\cdot}1F + 0{\cdot}9F^2 \tag{3.13}$$

$$\simeq 0{\cdot}3F + 0{\cdot}7F^2 \tag{3.14}$$

## 3.3.4 *Heating effect*

The energy from electrical power loss is converted to other energy forms, almost entirely heat. This heat energy thus tends to increase the temperature of the associated electrical component. High temperatures can result in premature ageing of insulation, while excessive temperature can result in conductors or insulation melting, with dangerous situations possibly occurring. The heating characteristic of a component depends on its material and construction, and there may be considerable temperature differentials throughout any item of equipment. In a large multimaterial piece of equipment the speed of heating and cooling differs in the various components. The transformer, with different winding, core and tank metals, plus conductor insulation and insulating oil, is a good example of this.

In this section the simple case of conductor heating will be considered. The effect of heating on transformers and cables is discussed in Chapter 6.

At any point in time the heat flow into a conductor due to losses is balanced by heat emission from the conductor plus heat retained in the conductor, in accordance with:

$$P_l \, dt = mc \, d\theta + a\alpha\theta \, dt \tag{3.15}$$

where $P_l$ = heating power

$\quad m$ = mass of conductor

$\quad \theta$ = (conductor temperature) − (ambient temperature)

$\quad d\theta$ = temperature rise during time $dt$

$\quad a$ = surface area

$\quad c$ = specific heat

$\quad \alpha$ = heat emission constant

In the case of a short circuit the period involved is very short and the last term in eqn. 3.15 covering emission can be omitted, giving

$$\theta_{sc} \simeq (P_l t)/(mc) \tag{3.16}$$

Given that $P_l = I^2 R$, the final temperature just before the clearance of a short circuit can be taken as being proportional to the protection operating time and to the square of the fault current, assuming the resistance to be independent of the temperature over the short period of time involved.

In the case of a steady load current the term $(mc \, d\theta)$ of eqn. 3.15 describing the retention of heat, is zero. Thus the difference between the final conductor temperature and the ambient temperature is given by

$$\theta_f \simeq P_l/(a\alpha) \tag{3.17}$$

and this is also proportional to the square of the load current.

In load-change conditions the time function of the temperature is obtained by solving the complete eqn. 3.15. When studying cooling $P_l = 0$. The solution leads to attenuating exponential curves. Their approximate equations are:

heating: $\quad \theta_t = \theta_f (1 - e^{t/\tau})$ $\qquad\qquad$ (3.18)

cooling: $\quad \theta_t = \theta_f e^{-t/\tau}$ $\qquad\qquad\qquad$ (3.19)

where $\quad \theta_t$ = temperature rise at time $t$ from switching on or off

$\qquad\quad \theta_f$ = final temperature rise

$\qquad\quad \tau$ = time constant depending on material, size and shape of the conductor, given by

$$\tau = mc/a\alpha \tag{3.20}$$

The general shapes of the heating and cooling functions are illustrated in Figure 3.4.

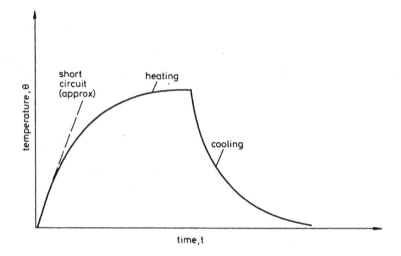

*Figure 3.4    Heating and cooling curves for a conductor*

## 3.4 Voltage drop

One of the most important constraints on distribution system design is the voltage level at the customer intake point. This is particularly important for the vast majority of customers taking supplies at low voltage with no means of adjusting the voltage received. A knowledge of the voltage at different locations can indicate the strong and weak parts of a network, and this is discussed in Chapter 13.

The voltage-drop phasor $V_d$ for a section of line having an impedance $Z$ and carrying current $I$ is given by

$$V_d = IZ \tag{3.21}$$

In distribution systems it is the arithmetic difference between the sending- and receiving-end voltages which is the more useful voltage-drop value. A close approximation to this can be obtained from the simplified equivalent circuit shown in Figure 3.5. The circuit has resistance $R$, reactance $X$, sending-end voltage $V_s$ and receiving-end voltage $V_r$. It carries current $I$ lagging on $V_r$. The equivalent phasor diagram is given in Figure 3.6

During normal load-flow conditions the angle between the receiving- and sending-end voltages $V_r$ and $V_s$ is only a few degrees. For most practical cases the approximation $\phi \simeq \phi'$ is acceptable, so that the scalar relationship can be written as

$$V_s = V_r + IR\cos\phi + IX\sin\phi \tag{3.22}$$

Figure 3.5    *Single-phase equivalent circuit for a section of line*

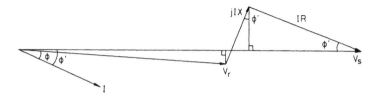

Figure 3.6    *Phasor diagram for the line represented in Figure 3.5*

The voltage drop $V_d$ in the line is given by

$$V_d = |\mathbf{V}_s| - |\mathbf{V}_r|$$
$$\simeq IR \cos \phi + IX \sin \phi$$
$$= I_p R + I_q X \tag{3.23}$$

In eqn. 3.23 $I_p$ and $I_q$ represent the resistive and reactive components of the load current $\mathbf{I}$. In single-phase calculations the resistance and reactance of the return path must be included in $R$ and $X$. For 3-phase systems the line–line voltage drop can be calculated from

$$V_d \simeq \sqrt{3}(I_p R + I_q X) \tag{3.24}$$
$$= \frac{P}{V}(R + X \tan \phi)$$

where $V$ is the line–line voltage and $P$ is the total three-phase power.

While eqns. 3.22–3.24 have been defined for flows along a line they are also appropriate for determining the voltage drop through any item of equipment in a network, knowing the equipment resistance and reactance and the load current and power factor. Referring back to the example of a line, the above study covered the case where the load was concentrated at the receiving end of a line section. Practical situations with several load points along the line can be solved by a series of similar calculations, obtaining the voltage drop due to each load point in turn. The total voltage drop is then the sum of the individual voltage drops due to each load point, assuming $\phi = \phi'$. If the loads do not peak at the same time it is necessary to apply a coincidence factor, defined in Section 3.3.1, to the loads to avoid obtaining too high a value for the voltage drop due to the simultaneous demand of the group of loads.

In some cases the load can be assumed to be distributed homogeneously along the line. In this situation the resultant voltage drop along the line will be one-half of the voltage drop obtained with a load equal to the total loading and

On HV/MV transformers the appropriate earthing arrangements can be applied to both the primary and secondary windings. On distribution systems it is only necessary to earth at the source supply point. Thus MV system earthing, if any, is normally carried out on the secondary neutral of the infeed HV/MV transformers. In addition, the metal frames of all transformers and other equipment are solidly earthed in HV and MV systems, whichever neutral earthing practice is adopted.

When applying methods (i), (ii) and (iii) the appropriate earth connection is made on to the neutral point of the transformer winding, e.g. at the neutral point of the star windings of a transformer, where the neutral of the network is directly available. Various earthing arrangements are shown in Figure 3.18 (page 57), which also shows the current distribution in the transformer windings under earth-fault conditions. With a delta winding an auxiliary transformer with star-interstar or zigzag windings is connected to the terminals of the delta winding to provide a neutral point. Consideration of the current flows for the various windings and earthing combinations is covered in Section 3.7.

### 3.5.4  Direct earthing

HV systems are often directly earthed. The earth-fault current may then exceed the 3-phase fault current. If not all of the neutrals of an HV system are earthed, the earth-fault currents and hazard voltages can be reduced to acceptable levels. In a directly earthed system power-frequency phase–earth overvoltages are the lowest, typically below 1·4 p.u.

### 3.5.5  Impedance earthing

MV networks often use different types of impedance earthing. Impedance earthing involves connecting a resistor or reactor between the system neutral point and earth. When a resistor is used its resistance is often of such a value that the earth-fault current passing through the transformer windings is limited to the rating of each winding. With an impedance earthed system the phase–earth voltage of an unfaulted phase can reach $\sqrt{3}$ times the normal value under earth-fault conditions, and occasionally some 5% higher, depending on the system $R/X$ ratios. This should not cause problems with system equipment since the insulation level in MV systems is based on much higher lightning overvoltages. Further consideration of system overvoltages is given in Section 3.8.

### 3.5.6  Arc-suppression-coil earthing

A specific example of a neutral earthing reactor is the arc-suppression, or Petersen, coil, whose inductance can be adjusted to match closely the network phase–earth capacitances, depending on the system configuration, so that the resultant earth-fault current is small. Thus the resultant touch or step voltage is small, so that most systems could be operated for long periods with a sustained fault until the fault can be cleared. The arrangement is shown in Figure 3.22.

Experience with MV systems has shown that it is possible to depart from the ideal tuning value by about 25% before operational problems with protection and high fault current appear. Instead of one large controlled coil at the HV/MV substation, in rural networks it is possible to place inexpensive small compensation equipments, each comprising a star-point transformer and arc-suppression coil with no automatic control, around the system. If these equipments are properly located on individual feeders around the distribution network, no additional automatically operating arc-suppression-coil compensation is required. The disconnection of the compensation equipment when the associated feeder is isolated from the network ensures that the overall network compensation is retained regardless of the switching arrangements on the network. With this system the uncompensated residual current remains somewhat higher than in automatically tuned compensation systems, but experience has shown that arc-quenching is not substantially worsened and is operationally acceptable.

### 3.5.7 *Isolated neutral systems*

Operating a system with the neutral isolated results in low values of earth-fault current equal to the system capacitance current (see Figure 3.19, page 58). The voltage between faulted equipment and earth is small, which improves safety. For the same hazard voltage, relatively higher protective earthing resistances are acceptable compared with most other neutral earthing systems. On the other hand, transient and power-frequency overvoltages can be higher than those obtained, for example, with resistance earthed systems. Figure 3.8 gives the phasor diagram for an overvoltage condition. Here the phase–earth voltage $V_A$ has increased to $V_{Af}$ due to a single-phase fault on phase B.

### 3.5.8 *Low-voltage-system protective earthing*

At low voltage the earthing arrangements should be such that, on the occurrence of a fault on any appliance, the potential on any exposed conducting part likely to be touched by an individual should not reach a dangerous level. Many different practices are applied, especially in buildings.

When dealing with this subject the following IEC codes have been applied.
First letter   = the relationship of the power system to earth:
     T   = direct connection of one point to earth,
     I   = all live parts, including the star point, isolated from earth, or one point connected to earth through an impedance.

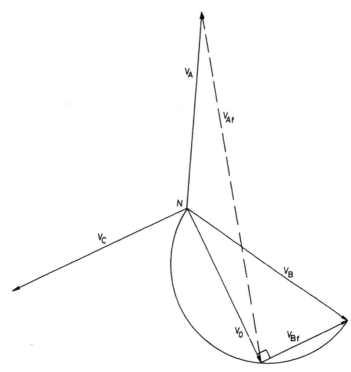

*Figure 3.8    Phasor diagram for a power-frequency overvoltage*
$$V_{Bf} = I_f R_f$$

Second letter = relationship of the exposed conductive parts of the electrical
installation to earth:

T = direct electrical connection of the exposed conductive parts to
earth, independently of the earthing of any point of the power
system,

N = direct electrical connection of the exposed conductive parts to
the earthed point of the power system (in AC systems the
earthed point is normally the neutral point).

Where neutral and protective functions are provided by separate conductors the
reference S is used. Where these functions are combined in a single conductor,
referred to as a PEN conductor, the code C is used.

In Figure 3.9a the equipment frameworks are connected to the neutral
conductor, which also includes the protective function, thus resulting in the
coding TN-C. It is usual to bond the metallic water and gas pipes within the
customer's installation to the earth connection, as well as the conducting frame
of any appliance, to prevent excessive potential differences occurring between
the non-electrical equipment and the earth system. A drawback of the TN-C
system is that the load current flowing through the PEN conductor causes
potential differences between the parts of the installation connected to the PEN

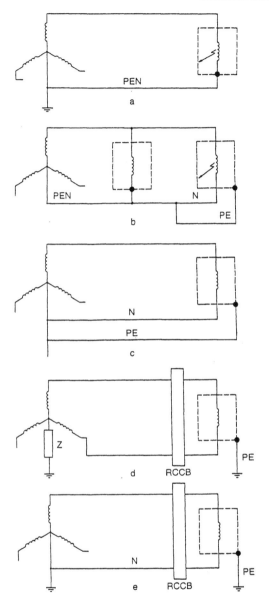

*Figure 3.9*  *Methods of earthing low-voltage systems*

RCCB: residual-current circuit breaker

$Z$: a high, or infinite, impedance

*a* Common neutral (N) and protective earth (PE) conductor (TN-C)

*b* Separate neutral and protective earth conductor in part of the system (TN-C-S)

*c* Separate neutral and protective earth conductors (TN-S)

*d* No direct connection between neutral and earth; separate protective earth electrode (IT)

*e* Electrically independent neutral and protective earth electrodes (TT)

Very often two assumptions can be made to simplify the calculations. It can be assumed that the pre-fault voltage at the point of fault is the same as the nominal system voltage at that point, and also that any load currents circulating around the system can be neglected since they will be small in comparison with the size of the fault currents.

Referring to Figure 3.11, the fault current of a 3-phase fault can be calculated using Thévenin's theorem by the equation $I_f = V/(Z_{th} + Z_f)$. Here $V$ is the phase–earth voltage at the fault location prior to the fault, and $Z_{th}$ is the total impedance seen from the fault, based on a single-phase equivalent network and including generator and motor impedances as well as circuit and equipment impedances. $Z_f$ is the fault impedance. In the simpler distribution systems the reduction of the network to determine $Z_{th}$ can often be carried out using simple network transformations.

Reference has already been made in Section 3.2 to the need to convert network impedances to a common base so that they can then be combined directly for network modelling. When calculating symmetrical 3-phase faults on a network, an equivalent single-phase network representation can be used. In addition to series and parallel reductions, star/delta and delta/star transformations may be required to obtain the equivalent impedance between source and fault.

Referring to Figure 3.12, the following relationships can be obtained and utilised in network reductions:

$$Z_A = (Z_{AB}Z_{AC})/(Z_{AB} + Z_{AC} + Z_{BC}) \tag{3.25}$$

$$Z_B = (Z_{AB}Z_{BC})/(Z_{AB} + Z_{AC} + Z_{BC}) \tag{3.26}$$

$$Z_C = (Z_{AC}Z_{BC})/(Z_{AB} + Z_{AC} + Z_{BC}) \tag{3.27}$$

$$Z_{AB} = Z_A + Z_B + (Z_A Z_B)/Z_C \tag{3.28}$$

$$Z_{AC} = Z_A + Z_C + (Z_A Z_C)/Z_B \tag{3.29}$$

$$Z_{BC} = Z_B + Z_C + (Z_B Z_C)/Z_A \tag{3.30}$$

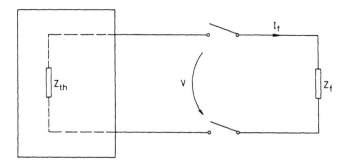

*Figure 3.11   Calculation of fault current*

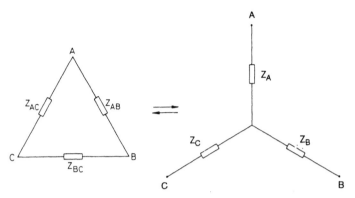

*Figure 3.12    Star/delta and delta/star transformation*

Figure 3.13 shows a small mesh network with two generation infeeds. It is required to find the fault level at point F. For simplicity only reactance values have been given, calculated as per-unit quantities on a 100 MVA base as described in Section 3.2. This is based on the assumption that circuit and generator resistances can be neglected since the generator reactances will predominate in any network reduction. The above assumption is used in the reduction of the reactances to a single value in order to calculate the fault level at F.

In order to calculate the fault level at F it is necessary to determine the total equivalent reactance between F and the reference node R, R being the point in the 3-phase system relative to which the phase voltages are expressed. When applying Thèvenin's method to this type of fault, R usually represents earth. R is also on of the notes for the sequence network used for building up the appropriate combination of sequence networks for unbalanced faults as described in Section 3.7. Thus, from the arrangement shown in Figure 3.13$a$, the equivalent network (Figure 3.13$b$) is obtained. Converting the delta reactances between busbars A, B and C to a star array $Z_A$, $Z_B$, $Z_C$, by use of eqns. 3.25–3.27, gives $Z_A = 0.022$, $Z_B = 0.017$ and $Z_C = 0.025$ as shown in Figure 3.13$c$. Combining and redrawing the reactances to the arrangement shown in Figure 3.13$d$, a further transformation of the delta reactances at the top of the diagram results in the network of Figure 3.13$e$.

Then follows a number of series/parallel reductions resulting in an overall source-fault reactance of 0.246 p.u. on 100 MVA as shown in Figure 3.13$h$. Since the reactances have all been calculated on a 100 MVA base an overall source-fault reactance of 1.0 p.u. would result in a 3-phase fault level of 100 MVA. With the calculated source-fault reactance for this example of 0.246 p.u. the resultant 3-phase fault level at point F is thus $100(1.0/0.246)$ $= 406.5$ MVA. If the network nominal system voltage were 33 kV, the fault level of 406.5 MVA would be equivalent to a fault current of $(406.5 \times 10^6)/\{\sqrt{3} \times 33000\} = 7.11$ kA.

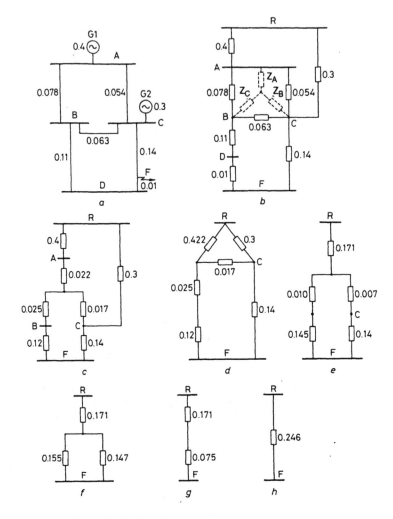

*Figure 3.13    Small mesh-network reduction example*

## 3.7 Unsymmetrical faults

### 3.7.1 Sequence networks

The simple single-phase equivalent circuit representation used in the previous section cannot be used directly when considering unsymmetrical faults since similar conditions do not apply to all phases of the faulted power system. Different methods of solving such problems have been developed, the most popular being the symmetrical-component system. This method is derived from an arrangement of three separate 'phase sequence' networks which, when suitably combined, provide the conditions for the unbalanced fault situation under consideration.

The three sequence networks do not actually exist within the power system but are purely a mathematical device to permit an easier solution of unbalanced system conditions. The detailed background to this method is given in many textbooks, and reference to some of these is included in the bibliography (Section 3.9). The use of symmetrical components is based on the fact that any combination of unbalanced 3-phase voltages and currents can be broken down into three separate systems of symmetrical phasors as illustrated in Figure 3.14 and comprising:

(i)   a positive-sequence system made up of three phasors having the same magnitude, spaced 120° apart and rotating in the same direction as the phasors for generated voltages in the power system under consideration, i.e. the positive direction;

(ii)  a negative-sequence system with three phasors of the same size spaced 120° apart, rotating in the same direction as the positive-sequence phasors, but in the reverse sequence;

(iii) a zero-sequence system where all three phasors are of the same magnitude and in phase with each other, rotating in the same direction as the positive sequence phasors.

Since each phase value is the sum of its sequence components, then, taking current values as an example,

$$I_a = I_{a1} + I_{a2} + I_{a0} \tag{3.31}$$
$$I_b = I_{b1} + I_{b2} + I_{b0} \tag{3.32}$$
$$I_c = I_{c1} + I_{c2} + I_{c0} \tag{3.33}$$

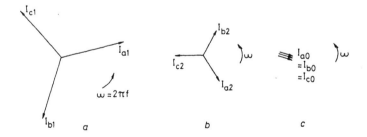

*Figure 3.14    Sequence-component phasors*

     *a*  Positive sequence
     *b*  Negative sequence
     *c*  Zero sequence

If $Z_{fn}$ is the sum of the phase–earth fault impedance $Z_f$ and the neutral–earth impedance then eqn. 3.43 is modified to

$$I_a = \frac{3E}{Z_1 + Z_2 + (Z_0 + 3Z_{fn})}$$

In a similar manner the following formulas for symmetrical 3-phase, line–line, and line–line–earth faults can be determined.

*3-phase fault*

$$I_a = \frac{E}{Z_1 + Z_f} \tag{3.44}$$

Phase currents $I_b$ and $I_c$ can be obtained from $I_a$ by multiplying by $a^2$ and $a$, respectively.

*Line–line fault*

$$I_a = 0; \quad I_b = \frac{-j\sqrt{3}E}{Z_1 + Z_2 + Z_f} \tag{3.45}$$

$$I_c = -I_b$$

*Line–line–earth fault*

$$I_a = 0$$

$$I_b = \frac{j\sqrt{3}(aZ_2 - Z_0 - 3Z_{fn})E}{Z_1 Z_2 + (Z_0 + 3Z_{fn})(Z_1 + Z_2)} \tag{3.46}$$

$$I_c = \frac{-j\sqrt{3}(a^2 Z_2 - Z_0 - 3Z_{fn})E}{Z_1 Z_2 + (Z_0 + 3Z_{fn})(Z_1 + Z_2)} \tag{3.47}$$

The interrelation of the phase-sequence networks for various system fault conditions is shown in Figure 3.17.

If the 3-phase fault current is assumed to have a value of 1 p.u. then the equations for the other types of fault can be defined as per-unit values of the 3-phase fault current at the same point, as set out below. Here $Z_f$ and $Z_{fn}$ are assumed to be zero:

Line–line fault current $= \sqrt{3}/(1 + Z_2/Z_1)$ per unit

Line–earth fault current $= 3/(1 + (Z_2 + Z_0)/Z_1)$ per unit

Line–line–earth fault current $= \sqrt{3}(aZ_2/Z_0 - 1)/(1 + Z_2/Z_1 + Z_2/Z_0)$ per unit

For lines and cables the positive- and negative-sequence impedances are equal. Thus, on the basis that the generator impedances are not significant in most

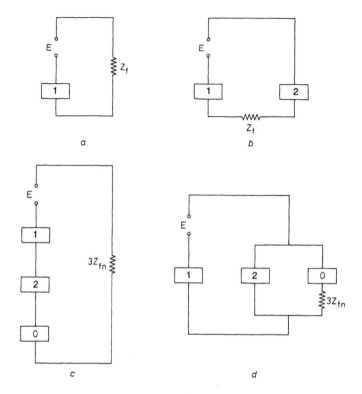

*Figure 3.17*  *Connection of sequence networks*

    *a*  3-phase fault
    *b*  Line–line fault
    *c*  Line–earth fault
    *d*  Line–line–earth fault

distribution-network fault studies, it may be assumed for simplifying the calculations that overall $Z_2 = Z_1$. From the above formulas the fault current for a line–line fault is $\sqrt{3}/2$ times that for a 3-phase fault at the same point on the network. The relationship for a line–earth fault reduces to $3/(2 + Z_0/Z_1)$, and to $\sqrt{3}(aZ_1 - Z_0)/(2Z_0 + Z_1)$ for a line–line–earth fault.

## 3.7.2 System sequence impedances

Typical parameters for overhead lines and underground cables are given in Table 3.1 (page 28). The zero-sequence impedance of an overhead line depends on the presence, or otherwise, of a neutral conductor or any earth wire. The zero-sequence reactance of underground cables varies according to the spacing between the conductors, and between each conductor and any metallic sheath or screen.

As the resistive component of large machines is usually so small, the impedances of synchronous machines, including generators, are often quoted as reactance values. The negative-sequence reactance is about 15–25% lower than the positive-sequence value, and approximately equals the subtransient reactance $X_{st}$. The zero-sequence reactance depends on the winding arrangement and no zero-sequence currents can flow unless at least one generator neutral is earthed in some manner. Typical generator parameters are given in Table 3.2 (page 29).

The transformer zero-sequence impedance is dependent on whether or not the winding arrangements and neutral earthing connections will permit the flow of balancing zero-sequence currents in the windings. Figure 3.18 shows a number of transformer winding arrangements, and the current flows through the windings on the occurrence of a phase–earth fault on the primary or secondary system. Figure 3.18a represents a star/star transformer with primary and secondary star points earthed. For an earth fault on the secondary winding, zero-sequence currents are free to flow in both windings. In the delta/star arrangement shown in Figure 3.18b zero-sequence currents can circulate in the primary delta winding, balancing the ampere-turns produced by the zero-sequence currents in the secondary star winding.

Figure 3.18c covers the case of a star/star transformer with the secondary-winding neutral point earthed, and a delta tertiary winding. Here the tertiary windings provide a path for the circulation of zero-sequence current, and in theory this would not be possible if the delta tertiary winding were absent. In this case the transformer presents a high reactance to the zero-sequence current, of the order of the magnetising reactance. The current will be higher for a 3-leg core-type transformer because of its higher zero-sequence leakage flux compared with a five-leg shell-type transformer.

In the star/delta arrangement in Figure 3.18d zero-sequence currents circulate in the delta secondary windings for a primary phase–earth fault, but do not escape into the secondary network. If the neutral point of the auxiliary transformer is connected to earth via an impedance $Z_g$ to limit the earth-fault current on the secondary side, the equivalent zero-sequence impedance of the neutral earthing impedance is $3Z_g$.

## 3.7.3 Earth faults

When considering earth faults it has so far been assumed that the network was earthed at some of the transformer neutral points, thus completing the earth-fault-current return path. With an unearthed system, or one earthed via an arc-suppression coil, the earth-fault current is small and often lower than the normal load current. Since this fault current is of the order of some tens of amperes it is unlikely to cause damage to lines, cables or other equipment.

Figure 3.19 represents an earth fault on an MV system having an isolated neutral. Here each line is modelled by a separate line–earth capacitance. Isolating all the neutral points on the system from earth causes the zero-phase-

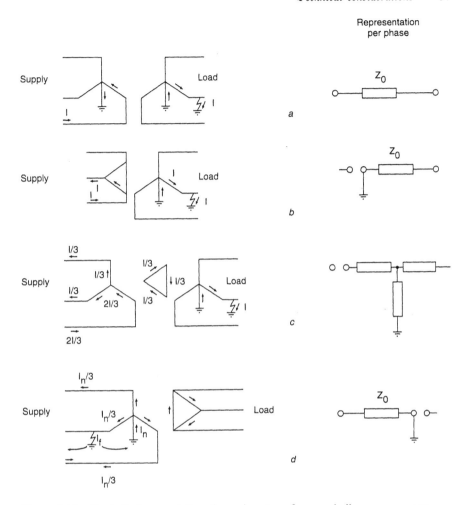

*Figure 3.18   Earth-fault current flow for various transformer-winding arrangements*

    *a* Zero-sequence currents free to flow in both primary and secondary circuits
    *b* Single-phase currents can circulate in the delta but not outside it
    *c* Tertiary winding provides path for zero-sequence currents
    *d* Single-phase currents can circulate in the delta but not outside it

sequence impedances between any point on the system and earth to appear as infinite. The series impedance of lines and equipment to zero-sequence current is essentially smaller than the shunt impedance represented by the earth capacitances of the lines, and can therefore be omitted. The earth-fault current path is completed via the line capacitances of each phase to earth.

Applying Thévenin's theory the circuit simplifies to Figure 3.20, with node *c* representing the neutral point of the MV winding of the HV/MV transformer. The line impedances have been omitted since they are small compared with that

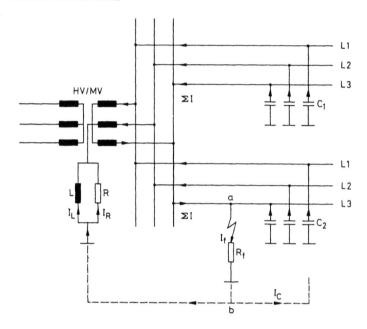

*Figure 3.22    Single-phase–earth fault on a system earthed via an arc-suppression coil*

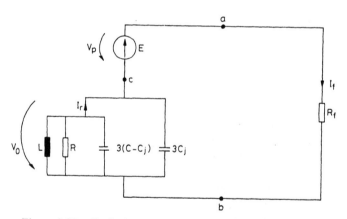

*Figure 3.23    Equivalent circuit for the arrangement in Figure 3.22*

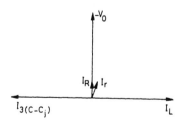

*Figure 3.24    Phasor diagram for the Petersen coil*

From the foregoing, voltage and current values and phase relationships can be calculated for different types of earth fault to assess stresses imposed on equipment, and for protection discrimination purposes, as in the following example.

*An example of a phase–earth fault on a system with isolated neutral compared with one on an earthed system*

In a phase–earth fault the three equivalent sequence networks are connected in series, as shown in Figure 3.16c. The magnitude of the zero-sequence impedance thus influences the nature of the earth-fault characteristics of the network. The zero-sequence network is the only one affected by the method of earthing.

The single-line diagram of the network used as an example is given in Figure 3.25. Both cases are calculated for this network, the only difference being that switch K is used to switch in or out the earthing resistor $R_y$. It is assumed that the 20 kV network supplied from the 110 kV substation is built up of lines having the same impedances and susceptances per unit length. The total length of lines is 200 km and the fault point $F$ is located on line 1 10 km from the substation. The relevant electrical parameters of the system are shown in the figure.

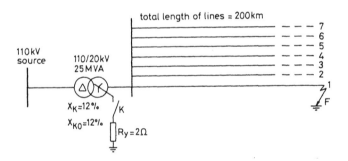

*Figure 3.25*   *Example network for phase–earth fault on an isolated, and an earthed, system*

    *a* Isolated system: switch K open
    *b* Earthed system: switch K closed
    Source data: $T' = 233$ ms $\Rightarrow X_1' = 4\cdot3\,\Omega$; $T_2 = 233$ ms $\Rightarrow X_2 = 4\cdot3\,\Omega$
    Fault point $F$ is 10 km from transformer
    Line data: $r = 0\cdot54\,\Omega/$km, $r_0 = 0\cdot69\,\Omega/$km, $x = 0\cdot38\,\Omega/$km
    $x_0 = 1\cdot52\,\Omega/$km, $b = 3\cdot0\,\mu$s/km, $b_0 = 1\cdot9\,\mu$s/km, and $I_1 = 10$ km

*(a) Isolated network*

Zero-sequence current cannot flow through the transformer since its neutral point is unearthed, so that calculation of the earth-fault is considerably simplified. In an isolated network there are no direct, or low-resistance, connections between the system neutrals and earth. This results in the zero-sequence impedances of the generators and transformers having infinite value, and the network zero-sequence impedance being determined by the earth capacitances of the lines. This capacitive impedance is some orders of magnitude higher than the positive- and zero-sequence impedances of the lines, which can therefore be neglected. The maximum value of the earth-fault current in the system being considered is thus virtually only dependent on the total capacitance to earth of the lines. This capacitance is proportional to the total line length and is much larger for underground cables than for overhead lines. The location of the fault does not have any significant effect on the magnitude of the fault current.

The earth-fault current in a phase–earth fault, from eqn. 3.48 is

$$I_f = \frac{V_p}{R_f + 1/(j\,3\omega C_0)} \tag{3.54}$$

where   $V_p$ = phase–earth voltage before the fault
$\omega = 2\pi f$ = angular frequency
$C_0$ = earth capacitance of one phase
$R_f$ = fault resistance

Setting $R_f = 0$ in eqn. 3.54, the maximum value of fault current is given by

$$I_{fmax} = 3\omega C_0 V_p = 3\omega c_0 l V_p = 3B_0 V_p = 3b_0 l V_p \tag{3.55}$$

where   $c_0$ = earth capacitance per kilometre of line
$l$   = total length of lines coupled together galvanically
$B_0 = \omega c_0 l = b_0 l$ = total earth susceptance of one phase
$b_0$ = earth susceptance per kilometre of line = $\omega c_0$.

Using the values given in Figure 3.25, from eqn. 3.55, the earth-fault current $I_{fmax}$ is

$$I_{fmax} = 3\left[\{(1.9 \times 10^{-6}) \times 200\} \times (20\,000/\sqrt{3})\right]$$
$$= 13 \cdot 2\,\text{A}$$

### (b) Earthed system

With switch K closed the earthing resistor $R_y$ provides a return path for the zero-sequence current. In this case the zero-sequence impedance of the transformer is typically about the same as its positive-sequence impedance. Figure 3.26 represents the connection of the three sequence-component networks in series for the phase–earth fault. In the zero-sequence network it will be seen that the resistor is represented as having a value of $3R_y$, and also that the equivalent transformer impedance is connected through that resistor to the reference point

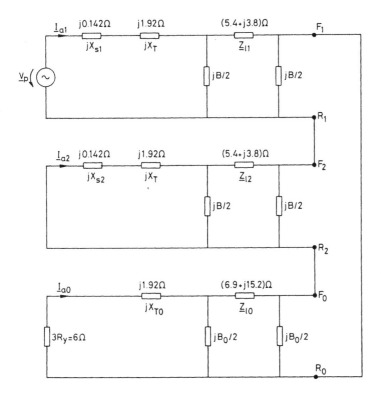

*Figure 3.26*  *Connection of sequence networks for an example of a phase–earth fault on a system with an earthed neutral*

$$\boldsymbol{Z}_{l1} = \boldsymbol{Z}_{l2} = I_1(r + jx); \quad B = I_l b;$$

$$X_{s1} = X_1' \left(\frac{V_2}{V_1}\right)^2; \quad X_{s2} = X_{s1}(X_2/X_1')$$

$$\boldsymbol{Z}_{l0} = I_l(r_0 + jx_0); \quad B_0 = I_l b_0; \quad X_{T0} = x_{K0}\frac{V_2^2}{S}$$

R. Thus a low-impedance path is provided for the circulation of zero-sequence current, and the impedances of the positive-, negative- and zero-sequence networks are of the same magnitude.

In Figure 3.26 the impedances have been expressed on the same voltage level, i.e. 20 kV. This network has been further simplified to produce the equivalent circuit shown in Figure 3.27 where the susceptances $B$ and $B_0$ have been neglected without affecting the accuracy of the result to any great extent. Since these susceptances are much smaller than that represented by the transformer earthed through the $2\,\Omega$ resistance, the assumption that $B = B_0 = 0$ opens the shunt branches of the equivalent circuit. By way of example, the susceptance $B/2$ corresponds to a reactance of 66 k$\Omega$. The parallel lines 2 to 7 (Figure 3.25)

value. The standard test-voltage lightning impulse is a full lightning impulse having a virtual front time of $1 \cdot 2\ \mu s$, and a virtual time to half value of $50\ s$, both within a given tolerance. If the surge voltage $V_1$ travels along a line with surge impedance $Z_1 = \sqrt{(L_1/C_1)}$, where $L_1$ and $C_1$ are the inductance and capacitance of that line, and arrives at a point where the surge impedance changes to $Z_2$, then the transmitted onward surge voltage $V_2$ is given by

$$V_2 = \frac{2Z_2}{Z_1 + Z_2} V_1 \tag{3.56}$$

The major increase in the transmitted surge is where $Z_2$ is considerably greater than $Z_1$. Thus if an overhead line feeds direct to a transformer having a surge impedance of several thousand ohms, then, from eqn. 3.56, the voltage surge introduced into the transformer will be almost twice the surge transmitted along the line. The appropriate method of protection is highly dependent on the local keraunic level, based on the average annual number of lightning strikes on a given area, which can vary by a factor of 100:1 for different regions of the world.

As well as externally produced overvoltages, transient or short-lived overvoltages can be produced within a network; for example due to a switching operation or a fault on the network. Owing to the system parameters, on MV systems such overvoltages may not reach the high values associated with lightning-induced overvoltages.

Earth faults may cause overvoltages between a healthy phase and earth. The maximum value of the earth-fault current $I_f$ for a system in which the neutral points are isolated is obtained from eqn. 3.48 as

$$I_f = -j\,3\omega C V_p \tag{3.57}$$

where $V_p$ is the highest phase–earth voltage.

The scalar value of the neutral-earth voltage $V_0$ is given by

$$V_0 = \frac{1}{\sqrt{\{1 + (3\omega C R_f)^2\}}} V_p \tag{3.58}$$

Thus

$$V_p^2 = V_0^2 + V_0^2 (3\omega C R_f)^2 \tag{3.59}$$

Eqn. 3.59 is the equation for a circle and, using eqns. 3.57 and 3.59, the phasor diagram shown in Figure 3.8 can be constructed. The Figure illustrates that, if the fault impedance is zero, the phase–earth voltage can have a value $\sqrt{3}$ times the normal voltage, and some fault-resistance values may even lead to slightly higher values of overvoltage.

In systems where the neutral points are earthed directly or via a small impedance overvoltages may occur during faults, but these are usually smaller than for the isolated neutral system described above. The maximum value of overvoltage depends on the ratios between different component resistances and reactances on the system as seen at the fault point. Figure 3.29 sets out

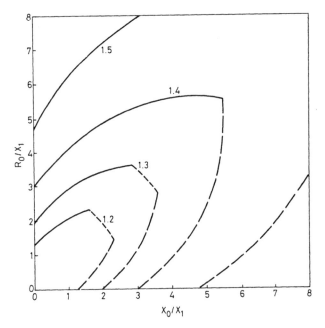

*Figure 3.29*    *Maximum value of the earth-fault coefficient $V/V_p$ with varying network parameters*

graphically the dependence of the maximum value of the *earth-fault coefficient* $V/V_p$ with varying network parameters.

In Figure 3.29 it has been assumed that the ratio between system resistances and reactances to the fault is given by $R_1 = R_2 = X_1$. The numbers on the curves indicate the maximum phase–earth voltage of any phase for any type of fault, in per-unit values of the nominal phase–earth voltage. In producing these curves the effect of fault resistance has been taken into account, using the value which results in the maximum voltage to earth. The discontinuity in the curves is caused by the effect of a change in the type of fault being assessed. The acceptable operating area, to avoid excessive high voltages being present on a network, is between the appropriate curve for the maximum acceptable per-unit ratio of $V/V_p$ and the $x$ and $y$ axes.

The individual resistances and reactances which influence the maximum overvoltage are associated with the type of earthing, the network components, the transformer primary–secondary winding arrangements and the network configurations. Networks where the earth-fault coefficient is lower than 1·4 are called 'effectively earthed'. With the assumptions of Figure 3.29 that requirement is met if $X_0/X_1 \leqslant 5·5$ and $R_0/X_1 \leqslant 5·5$. In partially earthed networks, where only some of the star points are earthed, the earth-fault coefficient is approximately 1·7, which means that the maximum phase–earth overvoltage can reach the level of the system phase–phase voltage.

Under certain resonant conditions internal overvoltages can be produced within a network. For normal system arrangements the system capacitances and inductances do not result in series resonance at 50 or 60 Hz. However, in exceptional fault conditions at time of low load, such resonance is possible. For example, when one phase conductor is broken, the line capacitances and the no-load impedance of a transformer may produce resonance at harmonic frequencies, resulting in overvoltages.

Further examples of resonant overvoltages are ferroresonance and jump resonance. Ferroresonance can occur when a transformer is energised via a lightly loaded, i.e. capacitive, circuit. The inrush magnetising current contains a series of harmonics, one of which may cause resonance between the circuit capacitance and the transformer inductance. For a voltage transformer connected to an unearthed system, the resonant circuit may be formed by the earth capacitance of the system and the non-linear inductance of the voltage transformer connected between phase and earth. Owing to the nonlinearity effect the inductance can operate at a lower impedance under certain conditions, giving resonant conditions at the basic frequency, or multiplies or subharmonics. At resonance the voltages and currents may jump from one state to the other, thus resulting in the term 'jump-resonance', and current overloading of the voltage transformer, leading to thermal damage, may occur. Modern voltage-transformer construction techniques and the use of suitable damping resistors effectively inhibit these types of overvoltages.

The various arrangements used to protect system equipment against damage from overvoltages are described in Section 7.6.

## 3.9 Bibliography

BAITCH, A.: 'Ferroresonance due to single-phase switching of distribution transformers'. International Conference on Electricity Distribution – CIRED 1979, AIM, Liege, paper 24

BHADRA, S. N.: 'A note on some computational aspects of the jump-resonance problem in 3-phase systems', *IEE Proc. C*, 1985, **132**, pp. 248–250

BLACKBURN, J. L.: 'Symmetrical components for power systems engineering' (Marcel Dekker, USA, 1993)

BRICE, C. W.: 'Voltage drop calculations and power-flow studies for rural electric distribution lines', *IEEE Trans.*, 1992, **IA–28**, pp. 774–781

CHARLTON, T., and GRIFFITHS, H.: 'High-voltage earthing system design and performance', *Power Eng. J.*, 1994, **8**, (4), pp. 173–181

CHEN, T.-H., CHEN, M. S., LEE, W.-J., KOTAS, P., and VAN OLINDA, P.: 'Distribution system short circuit analysis – a rigid approach', *IEEE Trans.*, 1992, **PS–7**, pp. 4444–4450

CLARKE, E.: 'Circuit analysis of AC power systems: Vol. 1 – Symmetrical and related components' (Wiley, 1943)

'Electrical installations of buildings: Part 3 – Assessment of general characteristics (Information on earthing systems)', IEC Publication 364–3, 1993

ERIKSSON, A. J., STRINGFELLOW, M. F., and MEAL, D. V.: 'Lightning induced overvoltages on overhead distribution lines', *IEEE Trans.*, 1982, **PAS–101**, pp. 960–968

FATHERS, D.: 'Use of PME earth terminals on overhead distributors', *Power Eng. J.*, 1994, **8**, pp. 261–264

HAASE, H., and TAIMISTO, S.: 'Distributed compensation of earth fault current in medium voltage networks', IEEE Conf. Publ. 250, 1985, 8th International Conference on Electricity Distribution, pp. 150–155

'High-voltage test techniques: Part 2 – general definitions and test requirements', IEC Publication 60–1, 1989

KOMARGI, K. G., and MUKHEDKAR, D.: 'Selected non-IEEE bibliography on grounding', *IEEE Trans.*, 1981, **PAS-100**, pp. 2993-3001

MAKINEN, A., and LAKERVI, E.: 'Engineering aspects for power system grounding in rural medium-voltage networks'. CIRED/UPEA Symposium on electricity distribution in the developing countries, Yamoussoukro, Ivory Coast, 1988

MELIOPOULOS, A. P., WEBB, R. P., and JOY, E. B.: 'Analysis of grounding systems', *IEEE Trans.*, 1981, **PAS-100**, (30), pp. 1039–1048

NAGRATH, I. J., and KOTHARI, D. P.: 'Modern power system analysis', (Tata McGraw–Hill, New Delhi, 2nd ed., 1989)

NAHMAN, J., and JELOVAC, D.: 'High-voltage/medium(low) voltage substation earthing systems', *IEE Proc. C*, 1987, **134**, pp. 75–80

PABLA, A. S.: 'Electric power distribution systems' (Tata McGraw–Hill Publishing Co., New Delhi, 1981)

PAUL, C. R.: 'System losses in a metropolitan utility network', *Power Eng. J.*, Sept. 1987, pp. 305–307

'Recommended practice for power system analysis' (IEEE Brown Book) ANSI/IEEE 399-1980

SALMON, R. G.: 'Neutral earthing reactors for 11 kV systems', *Distrib. Dev.*, Sept. 1975, pp. 15–19

SHIPLEY, R. B.: 'Introduction to matrices and power systems' (John Wiley, 1976)

STEVENSON, W. D., Jr.: 'Elements of power system analysis'. (McGraw–Hill, Tokyo, 4th ed., 1982)

SUMNER, J. H.: 'The theory and operation of Petersen coils', *Proc. IEE*, 1946, **94**, Part II, pp. 283–298

'Surge arresters: Part 1 – Non-linear resistor type gapped surge arresters for a.c. systems', IEC Publication 99–1, 1991

'Surge arresters: Part 4 – Metal-oxide surge arresters without gaps for a.c. systems', IEC Publication 99–4, 1991

TAGG, G. F.: 'Multiple-driven-rod earth connections', *IEE Proc. C*, 1980, **127**, pp. 240–247

WAGNER, C. F., and EVANS, R. D.: 'Symmetrical components applied to the analysis of unbalanced electrical circuits' (McGraw–Hill, 1933; reprinted 1982)

WEEDY, B. M.: 'Electric power systems' (Wiley, 3rd ed. 1979)

WILLHEIM, R., and WATERS, M.: 'Neutral grounding in high-voltage transmission' (Elsevier, 1956)

YUNG, E. K. N.: 'An innovative analysis of earthing grids for power substations', *IEE Proc. C*, 1985, **132**, (5), pp. 251–256

system. This would require that the summation of the associated costs, i.e. $C = \sum(C_i + C_l + C_m + C_o)$, should be minimised when planning the reinforcement scheme, where suffixes $i$, $l$, $m$ and $o$ refer to the costs of investments, losses, operation and maintenance, and outages, respectively.

An alternative approach is to determine a cost/benefit ratio, with the total costs associated with any project being calculated on an annual basis. These can be expressed as:

$$C = C_a + C_m - C_e \tag{4.1}$$

where  $C =$ total costs

$C_a =$ capital component of bringing forward the work by one year to achieve reduced NDE

$$= \frac{C_c \times (p/100)}{1 + p/100} \tag{4.2}$$

$p =$ annual interest rate

$C_m =$ annual cost of maintaining and operating the network extension

$C_e =$ reduction in the annual cost of system power and energy losses

$C_c =$ the capital cost of the project

If the annual amount of non-distributed energy is reduced by $W$ as a result of the network reinforcement, the cost/benefit ratio can be defined as $(C_a + C_m - C_e)/W$. This ratio has units of cost for each kWh improvement in non-distributed energy per annum. The lower its value the more justified is the proposed capital investment.

*Table 4.1    Calculation of cost of non-distributed energy per annum*

|  | Year 1 | Year 2 | Year 3 | Year 4 |
|---|---|---|---|---|
| 1 Non-distributed energy; present system, MWh | 5·05 | 10·48 | 15·00 | 17·20 |
| 2 Non-distributed energy; proposed system, MWh | 0·12 | 0·18 | 0·21 | 0·23 |
| 3 Saving in non-distributed energy = (1) − (2), MWh | 4·93 | 10·30 | 14·79 | 16·97 |
| 4 Annual capital charge, $C_a$, £ | 9090 | 9090 | 9090 | 9090 |
| 5 Operation & maintenance costs, $C_m$, £ | 2000 | 2000 | 2000 | 2000 |
| 6 Reduction in system losses, $C_e$, £ | 190 | 250 | 260 | 280 |
| 7 Net annual cost = (4) + (5) − (6), £ | 10900 | 10840 | 10830 | 10810 |
| 8 Cost per kWh saved = (7) ÷ {(3) × 1000}, £/kWh | 2·21 | 1·05 | 0·73 | 0·64 |

Table 4.1 summarises a cost/benefit exercise carried out to determine the cost per kWh saved. The capital cost of the proposed scheme is £100 000. From eqn. 4.2, using an annual interest rate of 10%, the annual capital charge $C_a$ is $(£100\,000 \times 0\cdot1)/(1 + 0\cdot1)$, say £9090. Operation and maintenance costs are assumed to be 2% of the capital cost. The non-distributed energy with the existing system is shown in line 1, with line 2 showing the lower value of NDE obtained with the proposed system reinforcement, resulting in the reduction in NDE given in line 3. The estimated saving in the cost of system losses with the proposed reinforcement is given in line 6. The annual cost of the scheme is therefore as shown in line 7, giving the cost per kWh saved as line 8.

It is necessary to adopt a threshold value for this cost/benefit ratio below which schemes are considered to be justified, and above which they are not justified. In the example in Table 4.1 if the threshold value was £1/kWh saved, the proposed reinforcement would hardly be justified in year 2 but would be well justified in year 3.

The value given to non-distributed energy, or 'energy not supplied', is not its cost, or the profit margin and interest charges included in its cost, but its value to the user when it is not available – in theory how much customers would pay in an emergency to have supply restored. Studies carried out internationally have shown that there are wide differences between the prices that individual customer groups put on the value of non-distributed energy, as shown in Figure 1.7. It is therefore difficult to assign a precise single threshold value for all customers. Various options are available, but in the end the threshold value will be a matter of judgment for any supply utility.

Additionally, such a cost/benefit assessment could take account of the savings obtained in kW not supplied, as well as the savings in energy not supplied, as considered in the example above. The cost/benefit equation could then be of the format $(C_a + C_m - C_e)/(aW + bP)$, where $P$ is the saving in kW not supplied and $a$ and $b$ are coefficients, their values being dependent on the importance customers place on both the amount of load lost and the length of time they are without supply.

For some specific customers having a low power demand for their appliances, e.g. a computer installation or a radio transmitter, the factors $a$ and $b$ may have extremely high values. In addition, when modelling the inconvenience due to very short outages, the energy involved is not a relevant factor.

## 4.3 Basic reliability theory

### 4.3.1 Components

Every component in a power system has an inherent risk of failure. In addition, outside factors influence the possibility of component failure; e.g. the current loading of the component, damage by third parties (human and animal), trees and atmospheric conditions – temperature, humidity, pollution, wind, rain, snow, ice, lightning and solar effect.

Given that $\lambda_i$ is the average outage rate, and $r_i$ the average outage duration of a component $i$, the expected annual outage time $U_i$ is given by

$$U_i = \lambda_i r_i \tag{4.3}$$

One of the main assumptions in developing the following theory and associated equations is that failures of individual components are independent of each other. When carrying out reliability calculations it is essential that the units used for $\lambda$, $r$ and $U$ are compatible. Thus, for example, $\lambda$ would be quoted in units of per annum, $r$ in hours and $U$ as hours per annum.

## 4.3.2 Radially operated systems

In the quantitative reliability analysis of a distribution system, the most important factors are the expected outage rates and times, and hence the outage demands and energies for individual customers and their total value for a specified network alternative. The latter implies a certain configuration and combination of network components based on a defined operation and maintenance policy.

An electrical circuit is composed of a number of components, such as lines, cables, transformers, circuit breakers, disconnectors etc. The solution is aimed primarily at estimating the influence of the unavailability of each component on the outages at each customer. Both the reliability data for the components used, and their location in relation to the paths from the feeding points to the customers, must be taken into account. For radially operated meshed networks, the relatively simple formulas in eqns. 4.4–4.7 can be used to calculate the reliability indices for a given load point $j$ when considering failures. In studies of medium voltage networks, a load point involves those customers fed by a particular distribution substation. The calculation parameters considered here are their average or expected values. In deriving these formulas it is also assumed that outage periods are much shorter than the mean time between outages.

The average outage rate or the number of outages per year for the load $j$ is given by

$$\lambda_j = \lambda_1 + \lambda_2 + \ldots \lambda_i + \ldots \lambda_n = \sum_{i \in I} \lambda_i \tag{4.4}$$

where $\lambda_i$ is the failure rate of component $i$ (per year), and $I$ is the set of the components whose failure results in an outage at the given load point $j$.

The expected annual outage time, sometimes referred to as 'unavailability', is given by

$$U_j = \sum_{i \in I} \lambda_i t_{ij} \tag{4.5}$$

where $t_{ij}$ is the outage time at the given load point $j$ caused by a failure of component $i$(h).

The average outage duration is then given by

$$r_j = U_j/\lambda_j \tag{4.6}$$

Outages occurring on a power system could result in some loss of the electrical energy being supplied. This load is usually termed 'energy not supplied' or 'non-distributed energy' (NDE), and is given by

$$E_j = \lambda_j r_j P_j \tag{4.7}$$

where $P_j$ is the outage power at the load point $j$.

The costs of the power and energy not supplied are given by

$$C_j = \sum_{i \in I} \lambda_i \{ a_j(t_{ij}) + b_j(t_{ij}) t_{ij} \} P_j \tag{4.8}$$

where $a_j(t_{ij})$ and $b_j(t_{ij})$ are the per-unit cost values for the demand and energy not supplied for the load point $j$ when the outage time is $t_{ij}$ (e.g. £/kW and £/kWh).

For permanent failures, the outage times $t_{ij}$ caused by a given component can include either equivalent switching times or repair time. At the end of the switching time the faulty component is isolated and the supply is restored to the given load point. The repair time is the time from failure until the supply is restored by repairing or by replacing the faulty component.

The equivalent switching time caused by any component depends on how the faulted component, the load point, network protection, locally and remote controlled disconnectors, and reserve connections are situated in relation to each other.

Seasonal variations and correlations between loads, failure rates, outage durations etc. can also be taken into the account by calculating each season separately or by considering respective correlation factors.

The formulas for temporary failures and maintenance outages are basically similar to those outlined above. Only the reliability and cost data are different.

### 4.3.3 Parallel or meshed systems

For two circuits or components in parallel, the outage rate due to overlapping outages on parallel circuits $\lambda_p$ is given by

$$\lambda_p = \lambda_1 U_2 + \lambda_2 U_1 \tag{4.9}$$

which from eqn. 4.3 can be written as

$$\lambda_p = \lambda_1 \lambda_2 (r_1 + r_2) \tag{4.10}$$

For two circuits or components in parallel the average outage duration $r_p$ is

$$r_p = \frac{r_1 r_2}{r_1 + r_2} \tag{4.11}$$

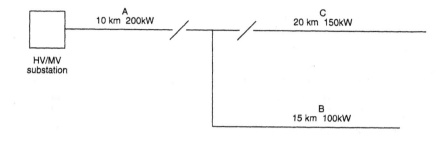

*Figure 4.1   Example network for reliability calculation*

*Table 4.1   Results for basic radial system (Figure 4.1a)*

| Load section | Fault section | $\lambda_i$ (1/year) | $t_{ij}$ (h) | $U_j$ (h/year) | $\lambda_i(a + bt_{ij})P$ (£/year) | $C$ (£/year) |
|---|---|---|---|---|---|---|
| A | A | 0·7 | 3 | 2·1 | $0{\cdot}7(0{\cdot}6 + 3 \times 3)200 =$ 1344 | |
| | B | 1·05 | 1 | 1·05 | $1{\cdot}05(0{\cdot}6 + 3 \times 1)200 =$ 756 | |
| | C | 1·4 | 1 | 1·4 | $1{\cdot}4(0{\cdot}6 + 3 \times 1)200 =$ 1008 | |
| | Subtotal | 3·15 | | 4·55 | | 3108 |
| B | A | 0·7 | 3 | 2·1 | $0{\cdot}7(0{\cdot}6 + 3 \times 3)100 =$ 672 | |
| | B | 1·05 | 3 | 3·15 | $1{\cdot}05(0{\cdot}6 + 3 \times 3)100 =$ 1008 | |
| | C | 1·4 | 1 | 1·4 | $1{\cdot}4(0{\cdot}6 + 3 \times 1)100 =$ 504 | |
| | Subtotal | 3·15 | | 6·65 | | 2187 |
| C | A | 0·7 | 3 | 2·1 | $0{\cdot}7(0{\cdot}6 + 3 \times 3)150 =$ 1008 | |
| | B | 1·05 | 3 | 3·15 | $1{\cdot}05(0{\cdot}6 + 3 \times 3)150 =$ 1512 | |
| | C | 1·4 | 3 | 4·2 | $1{\cdot}4(0{\cdot}6 + 3 \times 3)150 =$ 2016 | |
| | Subtotal | 3·15 | | 9·45 | | 4536 |
| | | | | | Total = | 9828 |

Table 4.1, from which it will be seen that the annual outage times for the three different zones are 4·55 h, 6·65 h and 9·45 h, with the total costs of outages being £9828 per annum.

Let us then consider the possibility of arranging backup connection. This produces the results shown in Table 4.2. In this case the reliability indices of sections B and C are improved while those for section A remain unchanged. The difference compared with the situation without backup connection is £1995/year.

*Table 4.2   Results for radial system with a backup connection (Figure 4.1b)*

| Load section | Fault section | $\lambda_i$ (1/year) | $t_{ij}$ (h) | $U_j$ (h/year) | $\lambda_i(a + bt_{ij})P$ (£/year) | $C$ (£/year) |
|---|---|---|---|---|---|---|
| A | A | 0·7 | 3 | 2·1 | $0·7(0·6 + 3 \times 3)200 =$ | 1344 |
|   | B | 1·05 | 1 | 1·05 | $1·05(0·6 + 3 \times 1)200 =$ | 756 |
|   | C | 1·4 | 1 | 1·4 | $1·4(0·6 + 3 \times 1)200 =$ | 1008 |
|   | Subtotal | 3·15 |   | 4·55 |   | 3108 |
| B | A | 0·7 | 1 | 0·7 | $0·7(0·6 + 3 \times 1)100 =$ | 252 |
|   | B | 1·05 | 3 | 3·15 | $1·05(0·6 + 3 \times 3)100 =$ | 1008 |
|   | C | 1·4 | 1 | 1·4 | $1·4(0·6 + 3 \times 1)100 =$ | 504 |
|   | Subtotal | 3·15 |   | 5·25 |   | 1767 |
| C | A | 0·7 | 1 | 0·7 | $0·7(0·6 + 3 \times 1)150 =$ | 378 |
|   | B | 1·05 | 1 | 1·05 | $1·05(0·6 + 3 \times 1)150 =$ | 567 |
|   | C | 1·4 | 3 | 4·2 | $1·4(0·6 + 3 \times 3)150 =$ | 2016 |
|   | Subtotal | 3·15 |   | 5·95 |   | 2961 |
|   |   |   |   |   | Total = | 7833 |

## 4.5 Bibliography

ALLAN, R. N., BARAZESH, B., and SUMAR, S.: 'Application of graphic/interactive computation in distribution system planning'. IEE Conf. Publ. 225, 1983, Reliability of Power Supply Systems, pp. 99–103

ALLAN, R. N., and BILLINTON, R.: 'TUTORIAL: Power system reliability and its assessment. Part 3 Distribution systems and economic considerations', *IEE Power, Eng. J.*, 1993, **7**, pp. 185–192

ALLAN, R. N., BILLINTON, R., BREIPOHL, A. M., and GRIGG, C. H.: 'Bibliography on the application of probability methods in power system reliability evaluation 1987–1991', *IEEE Trans.*, 1994, **PWRS–9**, (1), pp. 41–49

ALLAN, R. N., BILLINTON, R., and DE OLIVEIRA, M. F.: 'Reliability evaluation of electrical systems with switching actions', *Proc. IEE*, 1976, **123**, pp. 325–350

ALLAN, R. N., DIALYNAS, E. N., and HOMER, J. R.: 'Modelling and evaluating the reliability of distribution systems', *IEEE Trans.*, 1979, **PAS–98**, pp. 2181–2188

BOECKER, H., and KAUFMANN, W.: 'Optimising the expenditure on reliability of supply in system planning for the benefit of consumers', IEE Conf. Publ. 148, 1977, Conference on Reliability of Power Supply Systems, London

'Bibliography on the application of probability methods in power system reliability evaluation', *IEEE Trans.*, 1984, **PAS–103**, (2)

BILLINTON, R., and ALLAN, R. N.: 'Power system reliability in perspective', *Electron. Power*, March 1984, **30**, pp. 231–236

BILLINTON, R., and ALLAN, R. N.: 'Reliability evaluation of power systems' (Pitman, 1984)

BILLINTON, R., SATISH, J., and GOEL, L.: 'Hierarchical reliability evaluation in an electric power system', Proceedings of Joint International Power Conference 'Athens Power Tech' – APT'93, Athens, Greece, 1993, pp. 616–621

BILLINTON, R., and GOEL, L.: 'Overall adequacy assessment of an electric power system', *IEE Proc. C*, 1992, **139**, pp. 57–63

BILLINTON, R., and GROVER, M. S.: 'Reliability evaluation in transmission and distribution systems', *Proc. IEE*, 1975, **122**, pp. 517–523

BILLINTON, R., TOLLEFSON, G., and WACKER, G.: 'Assessment of electric services reliability worth', IEE Conf. Publ. 338, Third International Conference on Probabilistic Methods Applied to Electric Power Systems, PMAPS 91, 1991, pp. 9–14

BILLINGTON, R., WACKER, G., and SUBRAMANIAM, R. K.: 'Factors affecting the development of a residential customer damage function', *IEEE Trans.*, 1987, **PWRS–2** (1)

BROWN, W. C.: 'Cost versus reliability – a key to efficiency', *Trans. Distrib.*, 1982, 34, (2)

CABANE, E., CARTON, D., DENOBLE, R., GUILLEVIC, A., LATIL, L., and MICHACA, R.: 'Reliability of switchgear and transformers in distribution substations', IEE Conf. Publ. 99, 1973, International Conference on Electricity Distribution – CIRED 1973, pp. 36–42

EKWUE, A. O.: 'Economics and reliability of supply of developing power systems', *IEE Proc. C*, 1986, **133**, pp. 373–375

FISHER, A. G.: 'Practical reliability methods for 11kV and LV distribution systems, *Electron. Power*, 1982, **28**, pp. 504– 507

KJOLLE, G., and SAND, K.: 'RELRAD – an analytical approach for distribution system reliability assessment', *IEEE Trans.*, 1992, PWRD-7, pp. 809–814

KJOLLE, G., ROLFSENG, L., and DAHL, E.: 'The economic aspect of reliability in distribution system planning', *IEEE Trans.*, 1990, PWRD–5, pp. 1153–1157

KOLBAEK JENSEN, K.: 'Quality of supply. Evaluation of the costs of non-distributed energy', 7th International Conference on Electricity Distribution – CIRED 1983, AIM, Liege, Paper a16

KÄRKKÄINEN, S., VIROLAINEN, R., and VOLDHAUG, L.: 'Statistical distribution of reliability indices and unavailability costs in distribution networks and their use in the planning of networks', IEE Conf. Publ. 197, International Conference on Electricity Distribution – CIRED 1981, pp. 340– 343

MILLER, W. T., and DAWSON, D. G.: 'Cost of unreliability to consumers', *Transm. Distrib.*, 1982, **34**, (6)

MÄKINEN, A., and LAKERVI, E.: 'A flexible reliability assessment technique for power distribution networks'. 10th Power System Computation Conference, August 1990, Graz, Austria, pp. 809–816

MÄKINEN, A., PARTANEN, J., and LAKERVI, E.: 'Reliability evaluation as part of a computer aided electricity distribution network design', IASTED, International Conference on High Technology in the Power Industry, Phoenix, Arizona, USA, March 1988

MÄKINEN, A., PARTANEN, J., and LAKERVI, E.: 'A practical approach for estimating future outage costs in power distribution networks', *IEEE Trans.*, 1990, **PWRD–5**, (1), pp. 311–316

MIRANDA, V., ALMEIDA DO VALE, A., and CERVEIRA, A.: 'Optimal location of switching devices in a distribution system'. 7th International Conference on Electricity Distribution – CIRED 1983, AIM, Liege, Paper a13

O'CONNOR, P. D. T., and HARRIS, L. N.: 'Reliability prediction: a state-of-the-art review', *IEE Proc. A*, 1986, **133** (4)

PAPIC, M., and ALLAN, R. N.: 'Comparison of alternative techniques for the reliability assessment of distribution systems'. IEE Conf. Publ. 338, 1991, Third International Conference on Probabilistic Methods Applied to Electric Power Systems' PMPAS 91, pp. 174–179

STAIN, E.: 'The value of non-distributed energy as the development criterion of the power system'. EP/SEM.8/R.5. Geneva, UN Economic Commission for Europe, 1981

TOLLEFSON, R., BILLINTON, R., and WACKER, G.: 'Comprehensive bibliography on reliability worth and electrical service consumer interruption costs: 1980–1990', *IEEE Trans.*, 1991, **PS–6**, (4), pp. 1508–1514

WAUMANS, R. J. R., and BRUULSEMA, G.: 'How changing conditions affect design and reliability of distribution networks'. 7th International Conference on Electricity Distribution – CIRED 1983, AIM, Liege, Paper *a*12

*Table 5.1   Present-worth factors for various annual interest rates*

| Year | Annual interest rate, p% | | | | | |
|---|---|---|---|---|---|---|
|  | 5 | 7.5 | 10 | 12.5 | 15 | 20 |
| 0 | 1·000 | 1·000 | 1·000 | 1·000 | 1·000 | 1·000 |
| 1 | 0·952 | 0·930 | 0·909 | 0·889 | 0·870 | 0·833 |
| 2 | 0·907 | 0·865 | 0·826 | 0·790 | 0·756 | 0·694 |
| 3 | 0·864 | 0·805 | 0·751 | 0·702 | 0·658 | 0·579 |
| 4 | 0·823 | 0·749 | 0·683 | 0·642 | 0·572 | 0·482 |
| 5 | 0·784 | 0·697 | 0·621 | 0·555 | 0·497 | 0·402 |
| 6 | 0·746 | 0·648 | 0·564 | 0·493 | 0·432 | 0·335 |
| 7 | 0·711 | 0·603 | 0·513 | 0·438 | 0·376 | 0·279 |
| 8 | 0·677 | 0·561 | 0·467 | 0·390 | 0·327 | 0·233 |
| 9 | 0·645 | 0·522 | 0·424 | 0·346 | 0·284 | 0·194· |
| 10 | 0·614 | 0·485 | 0·386 | 0·308 | 0·247 | 0·162 |
| 11 | 0·585 | 0·451 | 0·350 | 0·274 | 0·215 | 0·135 |
| 12 | 0·557 | 0·420 | 0·319 | 0·243 | 0·187 | 0·112 |
| 13 | 0·530 | 0·391 | 0·290 | 0·216 | 0·163 | 0·093 |
| 14 | 0·505 | 0·363 | 0·263 | 0·192 | 0·141 | 0·078 |
| 15 | 0·481 | 0·338 | 0·239 | 0·171 | 0·123 | 0·065 |
| 16 | 0·458 | 0·314 | 0·218 | 0·152 | 0·107 | 0·054 |
| 17 | 0·436 | 0·292 | 0·198 | 0·135 | 0·093 | 0·045 |
| 18 | 0·416 | 0·272 | 0·180 | 0·120 | 0·081 | 0·038 |
| 19 | 0·396 | 0·253 | 0·164 | 0·107 | 0·070 | 0·031 |
| 20 | 0·377 | 0·235 | 0·149 | 0·095 | 0·061 | 0·026 |
| 21 | 0·359 | 0·219 | 0·135 | 0·084 | 0·053 | 0·022 |
| 22 | 0·342 | 0·204 | 0·123 | 0·075 | 0·046 | 0·018 |
| 23 | 0·326 | 0·189 | 0·112 | 0·067 | 0·040 | 0·015 |
| 24 | 0·310 | 0·176 | 0·102 | 0·059 | 0·035 | 0·013 |
| 25 | 0·295 | 0·164 | 0·092 | 0·053 | 0·030 | 0·010 |
| 30 | 0·231 | 0·114 | 0·057 | 0·029 | 0·015 | 0·004 |
| 35 | 0·181 | 0·080 | 0·036 | 0·016 | 0·008 | 0·002 |
| 40 | 0·142 | 0·055 | 0·022 | 0·009 | 0·004 | 0·001 |

option A is £9600 more expensive than option B. But since the majority of the net cash outflow in option B occurs in the earlier years, then on the present-worth basis, option A is £37410 cheaper and would be the preferred scheme. If the difference between any options were considered to be small, it might be necessary to look at the technical aspects of the options, possibly carry out a sensitivity analysis as described in Section 5.6, or consider other relevant factors as discussed in Section 5.8.

*Table 5.2    Present-worth example*

(a) Option A

| (1) Year | (2) Capital, £10³ | (3) Losses, £10³ | (4) Maintenance, £10³ | (5) Net cash flow, £10³ | (6) Present-worth factor $p = 5\%$ | (7) Present worth, £10³ |
|---|---|---|---|---|---|---|
| 0 | 0 | — | — | 0 | 1·000 | 0 |
| 1 | 0 | −1·0 | −0·5 | −1·50 | 0·952 | −1·43 |
| 2 | 0 | −1·2 | −0·75 | −1·95 | 0·907 | −1·77 |
| 3 | 0 | −1·4 | −0·6 | −2·00 | 0·864 | −1·73 |
| 4 | 0 | −1·6 | −0·6 | −2·20 | 0·823 | −1·81 |
| 5 | −10 | −1·9 | −0·8 | −12·70 | 0·784 | −9·96 |
| 6 | −40 | −2·1 | −1·0 | −43·10 | 0·746 | −32·15 |
| 7 | −70 | −2·4 | −1·0 | −73·40 | 0·711 | −52·19 |
| 8 | −60 | −2·6 | −0·8 | −63·40 | 0·677 | −42·92 |
| 9 | −20 | −1·5 | −0·7 | −22·20 | 0·645 | −14·32 |
| 10 | 0 | −0·8 | −0·4 | −1·20 | 0·614 | −0·74 |
| Totals | −200 | −16·5 | −7·15 | −223·65 | | −159·02 |

(b) Option B

| (1) Year | (2) Capital, £10³ | (3) Losses, £10³ | (4) Maintenance, £10³ | (5) Net cash flow, £10³ | (6) Present-worth factor $p = 5\%$ | (7) Present worth, £10³ |
|---|---|---|---|---|---|---|
| 0 | −20 | — | — | −20·00 | 1·000 | −20·00 |
| 1 | −75 | −1·0 | −0·5 | −76·50 | 0·952 | −72·83 |
| 2 | −80 | −1·2 | −0·6 | −81·80 | 0·907 | −74·19 |
| 3 | −20 | −1·4 | −0·6 | −22·00 | 0·864 | −19·01 |
| 4 | −5 | −1·05 | −0·5 | −6·55 | 0·823 | −5·39 |
| 5 | 0 | −0·8 | −0·4 | −1·20 | 0·784 | −0·94 |
| 6 | 0 | −0·8 | −0·4 | −1·20 | 0·746 | −0·90 |
| 7 | 0 | −0·8 | −0·4 | −1·20 | 0·711 | −0·85 |
| 8 | 0 | −0·8 | −0·4 | −1·20 | 0·677 | −0·81 |
| 9 | 0 | −0·8 | −0·4 | −1·20 | 0·645 | −0·77 |
| 10 | 0 | −0·8 | −0·4 | −1·20 | 0·614 | −0·74 |
| Totals | −200 | −9·45 | −4·6 | −214·05 | | −196·43 |

From Table 5.2 it will be noted that at the end of the review period the equipment installed under option B would be about five years older than that installed under option A. Where the present-worth values of the options are close, say within 10%, this difference in asset age can be taken into account by determining the *residual value* of the assets of each option at the end of the review period. These aspects are considered further in Section 5.4.

The annual cash flow can often be assumed to be made up of components which can be considered as constant over a number of years, incresing by a fixed percentage each year, or as increasing quadratically. These factors then modify eqn. 5.1 as follows.

If the annual cash flow is a constant $S_c$,

$$S_0 = S_c(100/p)\{1 - (1/\alpha^t)\} \tag{5.2}$$

where  $\alpha = (1 + p/100)$

$t$  = review period in years

When the load and costs increase by a fixed percentage $r\%$ each year,

$$S_0 = S_1\gamma(\gamma^t - 1)/(\gamma - 1) \tag{5.3}$$

where $S_1$ =costs in year 1 and

$$\gamma = (1 + r/100)^2/(1 + p/100)$$

Eqn. 5.2 was obtained from the sum of a geometric series with a multiplier for the ratio of successive terms of $1/\alpha$, and in eqn. 5.3 the multiplier used was $\gamma$. Eqn. 5.3 is particularly useful when modelling the costs of outages.

For the situation where costs have a quadrature relationship to the annual load growth $r$, the variable $\gamma$ is modified to $\gamma_1$, where $\gamma_1 = (1 + r/100)^2/(1 + p/100)$.

This modified version is suitable for loss calculations, particularly when considering the costs of the series, or copper, losses which have a quadrature relationship with loading.

## 5.3 Discounted cash flow

The example in Table 5.2 indicates one method used to compare schemes on an economic basis at a specified interest rate. However, in this case only capital expenditure and variations in the cost of losses and maintenance were considered. There was no reference to income, so that it was not possible to assess the *rate of return* on the capital employed in order to determine if the option with the lower present-worth value had an acceptable rate of return to the utility.

Both privately owned utilities and state supply authorities need to ensure that projects earn a *required rate of return* (RRR) in order to be considered economically worthwhile. The choice of minimum RRR is usually linked to the actual interest rate paid by a utility on capital borrowed and/or the degree of risk entailed in the project. State-supported authorities generally consider investment in distribution systems as low risk, and may therefore operate at RRRs around 5%. In this context it may be inappropriate to assess the rate of return of installing a supply to a single house, provided that the planning and design arrangements, and any capital contribution from the customer, meet criteria determined on a utility basis. Alternatively the rate of return may be calculated on the costs of providing supplies to a group of houses, again taking account of the expected income from sales of electricity plus the capital contribution from the housing-estate developer. On the other hand, schemes required to meet specific safety requirements, say, may not earn the RRR, and policy criteria on RRR must take account of such schemes.

An alternative method of assessing a project is to determine the *internal rate of return* (IRR). This is defined as the annual interest rate at which the discounted values of the cash inflows and outflows have the same gross present-worth value over a specified period; i.e. it is the discount rate at which the net present worth is equal to zero. It is normal to determine the IRR by an iterative process using various 'test discount rates' as in the following example.

Consider the situation where an investment of £6000 results in various financial benefits. These could be additional income, or savings in maintenance costs, or reduced system losses. In general, the benefits would be a mixture of these sums and the overall benefit for each year is shown under 'cash flow' in item (2) of Table 5.3 over the 10–year period. The present value of the benefits over the period are calculated using present-worth factors based on different test discount rates.

The computation of the calculations to determine the IRR of the above investment is set out in Table 5.3. Using the rules of present-worth practice the investment is allotted to year 0, and the annual benefits to years 1–10. The summation of the present worth of the benefits is compared with the present worth of the investment for each test discount rate. This is calculated as a ratio for expenditure/benefit. The IRR is the test discount rate when this cost/benefit ratio equals unity. The ratios from Table 5.3 are plotted graphically in Figure 5.1. The break-even point is at a test discount rate of 10·54%, which is thus the internal rate of return. A further column in Table 5.3 repeats the calculations at this rate to confirm the IRR. If the required rate of return is 8% then the expenditure of £6000, given the financial benefits quoted, is justified.

Such studies are amenable to desk-top-computer calculation. Subroutines can calculate the present-worth of salaries, system losses, maintenance work and income to arrive at a total present-worth of any option, including capital investment, enabling the IRR to be obtained. By deriving the IRR for various projects, then, if there is any limitation on the amount of capital investment in a particular year, i.e. capital rationing, optional projects can be placed in order of priority with preference being given to the projects with the higher rates of return.

divided evenly across the review period and added to the appropriate cash flows, the annual payment $S_a$ for a single investment $S_0$ is given by

$$S_a = S_0 \frac{p}{100} \quad \frac{1}{1 - \{1/(1 + p/100)^t\}} \tag{5.4}$$

Each method has its particular area of application. For example, when selecting the optimum size for a new feeder the present-worth method is suitable. On the other hand, the optimum year for replacing a conductor by one of larger cross-sectional area can be determined by comparing the annuity of the necessary investment with the annual savings in losses. When these are equivalent the optimum time has been reached provided that loadings do not decrease in the future.

An example involving capital investment and loss costs is indicated in Tables 5.4a and b. With the given load growth at one HV/MV substation, let it be assumed that there are two alternative methods of providing transformer capacity. The first is immediately to purchase a 16 MVA transformer, and add a similar-sized unit at a later date when necessary owing to load growth. The second method is immediately to purchase a 25 MVA unit which will meet the load over the 10 year study period.

The price of a 16 MVA unit is taken as £170 000, and a 25 MVA unit as £230 000. The no-load losses of the two transformers, $P_0$, are 16·1 and 21·8 kW, respectively, and the full-load series losses, $P_l$, are 88 and 121 kW. The cost of the no-load losses is taken as £170/kW, and the series losses £70/kW. The transformers are assumed to have a useful life of 20 years, and an 8% interest rate has been used in the present-worth and annuity calculations.

From eqn. 5.4, the factor by which a single investment should be multiplied to obtain the annuity can be calculated to be 0·102. By utilising the present-worth factor calculated from eqn. 5.1 the final columns in Tables 5.4a and b for the two strategies can be determined. In the first option the second transformer would have to be purchased in year 7. It will be noted that the difference in total costs is very small, so that any technical differences would dictate the final choice.

## 5.4 Time scale of studies and residual values

Electricity-distribution-system investment and operational costs are such that in practice the maximum period of assessment should be between 10 and 15 years. If any of the options being considered show similar rates of return, the period can be extended or a sensitivity analysis carried out. In these studies the capability of the various alternatives should be approximately the same at the end of the period being considered. If there is a significant difference in system capacity, some allowance should be made for this. This can be done by crediting any such scheme with a sum related to the value of the additional capacity, multiplied by the present-worth factor for the final year of the assessment period.

*Table 5.4    Summary of two alternative strategies for providing transformer capacity*

### a

| Year | Load, MVA | Losses, £/year $P_0$ | Losses, £/year $P_l$ | Annuity of investments, £/year | $\Sigma£$ $= (3) + (4) + (5)$ | Discounted present-worth value, £ |
|---|---|---|---|---|---|---|
| (1) | (2) | (3) | (4) | (5) | (6) | (7) |
| 1 | 11 | 2700 | 2900 | 17 340 | 22 940 | 21 240 |
| 2 | 12 | 2700 | 3500 | 17 340 | 23 540 | 20 170 |
| 3 | 13 | 2700 | 4100 | 17 340 | 24 140 | 19 170 |
| 4 | 14 | 2700 | 4900 | 17 340 | 24 940 | 18 330 |
| 5 | 15 | 2700 | 5600 | 17 340 | 25 640 | 17 460 |
| 6 | 16 | 2700 | 6300 | 17 340 | 26 340 | 16 590 |
| 7 | 17 | 5400 | 3500 | 34 680 | 43 580 | 25 410 |
| 8 | 18 | 5400 | 3900 | 34 680 | 43 980 | 23 750 |
| 9 | 19 | 5400 | 4300 | 34 680 | 44 380 | 22 190 |
| 10 | 20 | 5400 | 4800 | 34 680 | 44 880 | 20 780 |
| Total | | 37800 | 43800 | 242 760 | 324 360 | 205 090 |

### b

| Year | Load, MVA | Losses, £/year $P_0$ | Losses, £/year $P_l$ | Annuity of investments, £/year | $\Sigma£$ $= (3) + (4) + (5)$ | Discounted present-worth value, £ |
|---|---|---|---|---|---|---|
| (1) | (2) | (3) | (4) | (5) | (6) | (7) |
| 1 | 11 | 3700 | 1600 | 23 460 | 28 760 | 26 630 |
| 2 | 12 | 3700 | 1900 | 23 460 | 29 060 | 24 900 |
| 3 | 13 | 3700 | 2300 | 23 460 | 29 460 | 23 390 |
| 4 | 14 | 3700 | 2700 | 23 460 | 29 860 | 21 950 |
| 5 | 15 | 3700 | 3100 | 23 460 | 30 260 | 20 610 |
| 6 | 16 | 3700 | 3500 | 23 460 | 30 660 | 19 320 |
| 7 | 17 | 3700 | 3900 | 23 460 | 31 060 | 18 110 |
| 8 | 18 | 3700 | 4400 | 23 460 | 31 560 | 17 040 |
| 9 | 19 | 3700 | 4900 | 23 460 | 32 060 | 16 030 |
| 10 | 20 | 3700 | 5400 | 23 460 | 32 560 | 15 080 |
| Total | | 37000 | 33700 | 234 000 | 305 300 | 203 060 |

Since electrical equipment has a finite life owing to the normal process of deterioration, provision has to be made in the accountancy procedure of any organisation to 'write off' the equipment at the end of a given number of years, usually earlier than the expected average physical life. If before the end of this period it is necessary to remove equipment from the system because it has become obsolescent, or because of rapid technological development or changes in regulations, the equipment removed should be accorded a *residual value*. This should be the money which could be obtained from disposing of that asset even if the amount only reflects the value of the metal content of the equipment. An alternative approach is to include as a residual credit a discounted value for future cash inflows from the project.

Where an item of equipment is replaced, and the recovered equipment can be removed for use elsewhere on the system, the scheme covering the equipment replacement should be credited with the present-worth of the residual value of the released equipment at that time, having regard to the unexpired portion of its average life expectancy.

The present worth of the residual value can be determined in the same way as that of any single annual cost. The appropriate present-worth factors in Table 5.1 should be applied to residual values in order to obtain equivalent present values. It will be noted that the residual value of equipment reaching its accountancy age, usually between 25 and 40 years, can be quite small. However, where equipment is changed relatively soon after installation, the residual value can have a significant influence on the economic assessment.

## 5.5 Inflation and interest rates

The question of whether to include the effect of inflation in economic assessments has long been a matter of debate and conjecture. The simplistic view is that any rise in the costs of the resources for any project will generally keep in step with the average annual inflation rate over the period of years involved. When items of cost are fixed, e.g. via a maintenance contract or rental, such items should be discounted in order to include the cost at current price levels.

Equally the selling price of electricity, which mainly provides the income to finance capital investment projects, can be presumed to keep roughly in line with general price levels. By improving operational efficiency, the cost of electricity relative to other goods and services can, however, be reduced over a number of years, and this aspect should also be considered. Most capital-scheme investment appraisals cover a similar period of time. Given that the costs of the various options are comparable, then ignoring inflation will not significantly distort the economic comparisons provided that inflation remains at a modest level, say 5% per annum. Equally the vast majority of distribution network projects are mostly at MV/LV, so that economic appraisals over a period of five years or so are the most common. In this context inflation will only have a minor effect.

It should be recognised that periods of inflation at the 20–25% per annum level have occurred over the past 15 years, while inflation levels considerably in excess of 100% per annum have been experienced in some countries. On the assumption that all income and expenditure are equally affected by inflation, the effect of inflation will be to increase the present-worth in year $t$ by $(1 + a/100)^t$ where $a$ is the annual rate of inflation. Thus in year $t$ the effective present-worth factor then becomes $\{(1 + a/100)/(1 + p/100)\}^t$ where $p$ is the annual interest rate.

The relationship between the market interest rate of loans and inflation has an influence on the optimum RRR of the utility, and therefore on its investment policy. Inflation is the only factor which tends to reduce the required RRR below the interest rate. If inflation is relatively high, say 15%, and market interest moderate, say 10%, the real interest rate becomes negative. An RRR of a few percent can then be applied, so that it may be economic to borrow large sums of money for earlier capital investment. In low inflation conditions, say 2%, and with interest rates around 10% for example, there is then a fairly high positive real interest rate and the RRR has to exceed the interest rate to be economically viable. This tends to a much lower level of investment than in the previous example.

Where items of the investment appraisal calculation are being forecast to increase at differing levels of inflation, then the discounted cash flow can be carried out by using anticipated 'out-turn' prices. These prices should then be discounted at a rate which is equal to the sum of the required rate of return and the average inflation rate in order to eliminate the differing inflation rate. Such differing rates could apply to various fuels such as coal, gas etc., or costs related to the price of fuels, or salaries if these did not keep in line with inflation.

When utilities have to raise loan capital on the open money market, the minimum RRR, assuming negligible inflation, is influenced by the market interest rate. In such circumstances, the minimum RRR for economic studies needs to be a few per cent higher to provide a margin of profit and to acknowledge the uncertainties of forward forecasting. If optional capital is available, this should be invested where it can earn the highest return, with the return being used in economic studies. This situation is more applicable to developing countries particularly if capital-loan raising is on a national basis. On the other hand, state-supported utilities may be able to obtain capital at reduced interest rates.

## 5.6 Sensitivity analysis

The use of capital-investment appraisal techniques is intended to increase the confidence with which investment decisions are made. While there is no particular difficulty in using the techniques, it is more difficult to ensure that the forecast figures used are correct for all the possible changing circumstances during the economic life of a project. A sensitivity analysis measures the

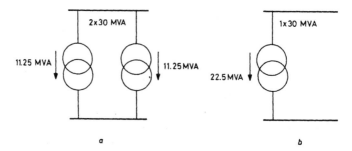

*Figure 5.2   Transformer-loss studies*

Annual average values for losses must not be used for operational studies since the variation of instantaneous energy prices is high. In these cases it is more appropriate to carry out a more detailed consideration of the different prices at different load levels. Such a situation is indicated in Figure 5.2 where consideration is being given to determining under which load conditions one transformer, or two, should be switched in and loaded. In high load conditions the cost of losses for two transformers will be lower than those for a single transformer. This is because the series losses vary as the square of the current so that the costs of the series losses with two transformers are one-half of those with one transformer and, at a particular load level, this more than offsets the higher cost of the shunt losses for the two units. Because the instantaneous values of losses, per kW, under high load conditions are typically high, this crossover point may be reached at considerably lower load levels than conventional calculations with average annual loss values might indicate.

## 5.8 Other factors

Economic comparisons relating to electricity-supply networks are usually confined to the financial and technical aspects of various options. It will probably also be necessary on occasion to take account of any additional costs involved in minimising the impact of any work on the amenity of an area. As a consequence of such constraints, the possibility of extra expenditure, such as diverting overhead lines or undergrounding sections of individual circuits, must be taken into account, either in the costing of the scheme or in a sensitivity analysis. It may well be that some other option, previously considered too expensive, may then prove more cost-effective than the originally preferred scheme plus the added amenity expenditure.

The question of state or municipal taxes and duties has not been covered, as these vary considerably from one country to another and generally act as an additional cost to be borne by the utility. Compensation for obtaining overhead line or underground cable routes will, however, form part of the costing exercise for any system extension proposal.

The social requirements of providing electricity supplies to remote areas may result in the investments not meeting the minimum RRR in some cases. Such schemes may even be installed at a loss or require financial supply from outside the supply authority from government, local communities or international sources.

It may also be necessary to carry out work on an electricity supply system to meet revised safety or technical regulations, and provision should always be made in any long-term financial plan for some such additional expenditure. The introduction of electricity supplies into an area can have additional spin-off in an improved economy not directly related to energy sales. All or some of the factors mentioned here, and other relevant ones, may need to be taken into account to arrive at a fair and reasonable economic assessment.

Notwithstanding all the procedures and aspects mentioned in this Chapter there will always be the overriding consideration that the supply utility must remain economically viable. This may lead to cutbacks in proposed projects in some years or the possibility of additional money becoming available as energy sales expand, and the planning engineer should always have these aspects in mind.

## 5.9 Bibliography

BERRIE, T. W.: 'Power system economics' (Peter Peregrinus, 1983)

BERRIE, T. W.: 'Electricity economics and planning' (Peter Peregrinus, 1992)

BINNS, D. F.: 'Economics of electrical power engineering' (Electrical Logic Power Ltd., PO Box 14, Manchester M16 7QA, 1986)

NICKEL, D. L., and BRAUNSTEIN, H. R.: 'Distribution transformer loss evaluation', *IEEE Trans.*, 1988, **PAS-100**, pp. 788–810

NORRIS, T. E.: 'Economic comparisons in planning for electricity supply', *Proc. IEE*, 1970, **117**, pp. 593–605

PERSOZ, H.: 'Financial constraints, discounting and growth', *Rev. Gén. Elect.*, April 1982, pp. 251–256

SAIED, M. H., and EL SHEWY, H. M.: 'Optimal design of power transformers on the base of minimum annual cost'. IEEE 1979 Joint Power Generation Conference, Charlotte, North Carolina, Oct. 1979

WISTBACKA, H.: 'Low loss versus initial value – Getting the best value', *Elect. Rev.*, 1952, **211** (18)

*Chapter 6*

# Equipment

## 6.1 General

In all other Chapters the objective has been to present the various theoretical, technical, economic and operational factors to be considered when planning and designing electrical distribution systems. This chapter is intended to provide some background information on the construction and operating characteristics of the main components installed on distribution networks, in order to provide the design engineer with basic data on the equipment which will be used to build up a functional system. With rapidly changing technologies affecting in some way the design of virtually all equipment, specific examples have not been illustrated; only general aspects of the main features of transformers, lines, cables and equipment are covered in this chapter. Protection aspects are covered in Chapter 7 and switchgear arrangements in Chapter 8.

From the commencement of manufacturing electrical equipment, various local and national standards have been used as the basis of individual equipment design. This has led to some non-uniformity of approach, and difficulties in reconciling equipment built to different specifications. The introduction of the International Electrotechnical Commission (IEC) standards provided a common base for the reliability and safety of equipment, interchangeability and mutual compatibility of equipment made by different manufacturers worldwide, plus the elimination of the unnecessary diversity of components used in the construction of electrical equipment. As a consequence, the IEC standards are used as the basis for regional and national standards, and in preparing specifications for international trade. They thus represent essential reference works for planning and design engineers.

Subject to network operating conditions permitting their retention, service experience has shown that equipment on electricity supply systems can have long engineering lives. HV/MV transformers are still giving good service over 50 years after being installed, as are MV cables, some of them now operating at twice their original design voltage. Examples are known of MV switchgear being replaced 90 years after installation and still in serviceable condition. The operating principles used in their construction were somewhat simple, using the traditional conductors, insulation and magnetic materials of that time. Many new applications have been developed using devices from outside the power-engineering sector, such as microelectronics, computer science, telecommunications and material technology – all well proven before being applied to power-distribution practice.

New equipment can, however, be more economic in operation with lower losses, better reliability and longer intervals between maintenance servicing; and these factors often motivate equipment replacement, as discussed in Section 6.6. In many cases replacement of equipment has been implemented because of changes in governmental or utility safety rules, or the lack of suitable spare parts as manufacturers have combined and rationalised their products, plus technological advances, rather than because of an inherent deterioration of ageing equipment.

## 6.2 Transformers

### 6.2.1 General

The transformer provides the facility for interlinking systems operating at different voltages. Thus voltage levels can be stepped up at power stations and transmission substations for supplying inter-regional power tie lines, and then reduced in stages through the HV, MV and LV networks in turn as indicated schematically in Figure 1.5. HV/MV and the larger MV/LV transformers are ground-mounted on concrete pads. Figure 6.1 shows an outline drawing of a typical HV/MV arrangement, identifying the major components. These transformers may also be located in 'indoor' installations within a suitable building for environmental reasons, particularly in urban areas.

In the so-called underground residential distribution (URD) system, which has found most favour in the USA, the MV/LV transformers are located in special underground enclosures. Smaller-sized units are often mounted on a platform constructed between two or four poles, with the smallest-sized transformers attached directly to a pole. Further reference to these pole-mounted transformers is given in Chapter 10.3. It is not uncommon for a cluster of three units to be attached to one pole, especially in the USA where single-phase units are preferred.

### 6.2.2 Losses

Transformer losses can be divided into two components: no-load losses and load losses. No-load losses result from hysteresis and eddy-current losses in the iron core, which depend on the type of steel used to fabricate the core. These losses are independent of the load current passing through the transformer, but increase with increasing voltage.

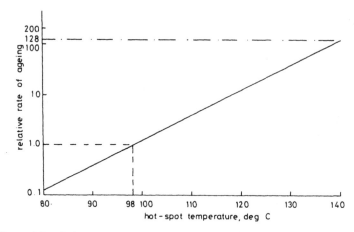

*Figure 6.2   Relative ageing of transformer insulation with hot-spot temperature.*

Figure 6.3 shows the amount of overload that can be carried by an ONAN transformer depending on the ambient temperature, and the loading cycle. Curve *c* covers the situation of continuous loading for 24 hours per day. It will be seen that full load can be carried at an ambient temperature of 20°C, rising to 1·3 times full load at −20°C and falling to 0·85 times full load at +40°C. Curve *d* shows the limit of loading which would lead to rapid deterioration and damage to the insulation if the load were applied continuously.

Curves *a* and *b* cover conditions where the maximum load occurs for 2 and 8 h continuously each day. Suffix 1 is where the previous loading is 0·25 times the transformer rating, while for suffix 2 the previous loading is 0·9 times the transformer rating. If high loadings are being considered, studies should be carried out to ensure that associated components such as the bushings and the transformer tap changer can carry the overloads in each daily load cycle. There are other factors besides ageing which affect the service life of a transformer, for example uprating the distribution voltage, increased loads, or reduced losses of new transformers, will mean that transformer replacement will generally occur well before the electrical or mechanical life span due to ageing has been reached.

## 6.2.4 Reactance of large supply transformers

The chosen design impedance of a transformer is basically determined by two factors which oppose each other. The first is to keep the fault level on the lower-voltage side to an acceptable value. This can result in somewhat high impedances which can then produce too high a voltage drop through the windings, so that a balance has to be struck between these two constraints. Linked to the voltage-drop aspect is the range of tappings provided on the transformer to adjust the winding ratio. On the assumption that the power flow is from the higher-voltage to the lower-voltage busbar, the tapping range must be such as to obtain an acceptable voltage at the lower-voltage busbar at any transformer loading for any likely variation in the higher-voltage busbar voltage.

It is possible that other considerations may restrict the maximum reactance. These include the need to minimise step change in voltage on the loss of one transformer, or on the loss of load, where two or more transformers may be operating in parallel on the lower-voltage side. In addition, the through-fault current must be adequate to operate protective relays, and the fault level at the lower-voltage busbar must be high enough to minimise the voltage elasticity $(dV/dP, dV/dQ)$ and thus the effect of disturbing loads. The maximum value of reactance should be derived for the most limiting condition. In determining the tapping range, only sufficient taps need to be provided to cover variations in the primary and secondary busbar voltages, and the loading on the transformer.

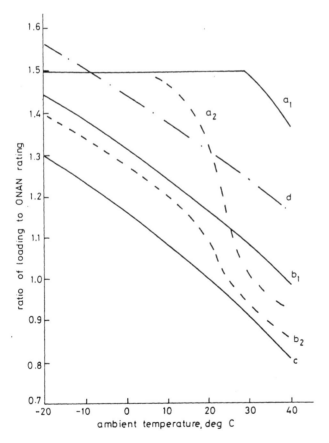

*Figure 6.3 Typical maximum loadings on an ONAN transformer, dependent on ambient temperature and previous loading*

### 6.2.5 Unsymmetrical loads

In low-voltage networks most customer appliances are single phase, and therefore tend not to be evenly distributed across the phases of a 3-phase supply. In addition, where only single-phase supplies are provided, this can result in unsymmetrical loadings. The smallest distribution transformers (MV/LV) in some countries are single-phase units. The primary side of the distribution transformer may be connected between phase and earth or across two phases of the MV line. Thus both voltage and current asymmetry can appear on 3-phase MV circuits.

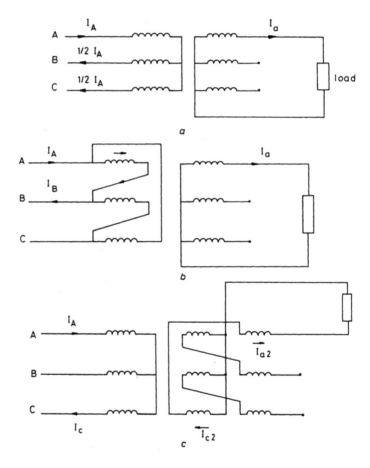

*Figure 6.4   Distribution of unbalanced load currents for different winding arrangements*

   *a*  Star/star
   *b*  Delta/star
   *c*  Star/zigzag

The degree of asymmetry transferred to the MV system depends also on the winding arrangements of the MV/LV transformers. Two types of winding connections are used in 3-phase distribution transformers: either the delta/wye (*Dy*) or the wye/zig-zag (*Yz*), shown in Figure 6.4.

If a 3-phase star/star-connected bank of three single-phase units has its primary star point connected to that of a *YNyn* supply, the load on each primary phase is proportional to that on the corresponding secondary phase, and the degree of unbalance is therefore the same on both sides of the transformer. When the primary star point is not connected, as in the *Yyn* arrangements shown in Figure 6.4a, the voltage of the primary neutral N approaches the voltage of phase A while voltages B–N and C–N increase. Since the voltage of the loaded phase is low, the result is that the secondary phase voltages are also unbalanced. To prevent this, star/star connections are not used in distribution transformers.

In order to avoid the above-mentioned voltage asymmetry the delta/star connection shown in Figure 6.4b can be used. The delta-connected primary will permit single-phase unbalanced secondary loading even though a primary neutral return is now absent. The secondary load current will cause a primary current flow from phase A to phase B.

Another suitable arrangement is the *Yz* connection shown in Figure 6.4c. The single-phase load current now flows through two sub-windings. This causes the MMFs on the different core limbs to maintain their balance so that no great voltage asymmetry will appear. The *Yz* connection is frequently used, especially in small transformers.

## 6.3 Overhead lines

An overhead line utilises air to insulate bare conductors for the majority of its length, apart from the very small sections connected to insulators on the line supports. The absence of further insulation and the relative ease of construction result in an inexpensive method of providing an electric circuit compared with an underground cable, particularly at MV and the higher voltages. Table 6.1 compares the costs of providing a circuit by overhead line or underground cable at different voltage levels. An appropriate average circuit rating, which is different for each of the four voltage levels, has been used for comparison purposes.

*Table 6.1    Ratio of underground-cable to overhead-line costs, at each voltage level*

| EHV | HV | MV(a) | MV(b) | LV(a) | LV(b) |
|---|---|---|---|---|---|
| 15 to 25:1 | 10:1 | 5:1 | 2:1 | 5:1 | 1.5:1 |

(a) in built up areas
(b) in normal soil conditions

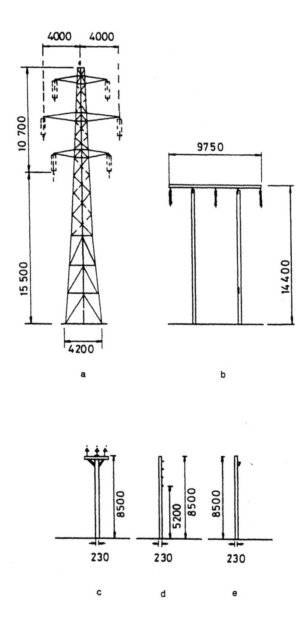

*Figure 6.5    Towers and poles for overhead lines*
Dimensions are in mm

For LV networks three main types of construction are in general use. Separately covered conductors are widely used for distributors and services in the UK and USA. Aerial bunched cables are used in Finland and France, for example, while self-supported aerial cables are in common use in Italy. Other constructions involve an insulated neutral, rather than the neutral wire being bare. A typical LV aerial bunched cable is shown in Figure 6.6. Experience in Nordic and tropical countries has shown that minimal tree trimming is necessary to erect and operate aerial bunched cables.

Similar types of insulated line are also used in MV systems. At medium voltage the insulation costs are much higher, so that the last two constructions mentioned in the previous paragraph are only economic under special circumstances. A 20 kV double-circuit line with covered conductors is shown in Figure 6.7. In this type of arrangement the conductor insulation does not meet IEC standards for full line or cable insulation, but it is nevertheless satisfactory for preventing a power fault occurring with clashing conductors or when conductors strike trees in high winds. The conductor insulators, however, have to cope with the same voltage stresses as occur with bare overhead conductors. At medium voltage, covered conductors provide a more economical arrangement than overhead cables.

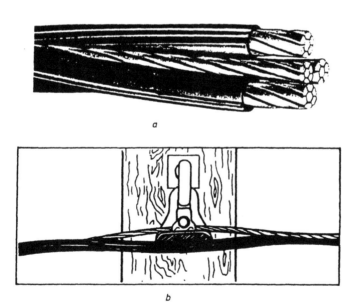

*a*

*b*

*Figure 6.6   Typical LV aerial bunched cable*

    *a* Section of self-supporting overhead cable
    *b* Cable suspended from a pole.

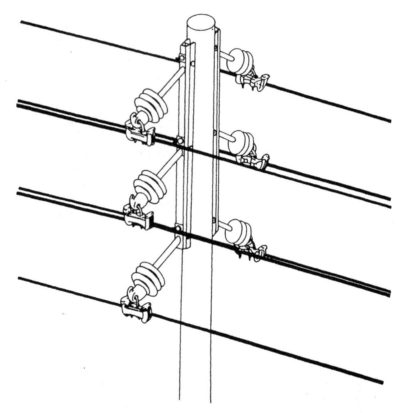

*Figure 6.7   20 kV double-circuit line with covered conductors.*

## 6.4 Underground cables

When supported by insulators from poles or towers an electrical conductor can utilise the air surrounding it as the insulating medium. In placing a conductor underground, buried directly, or in ducts or pipes, the first consideration is to insulate each conductor from the other conductors, and from earth (ground).

An early form of cable-conductor insulation, introduced in the late 1880s, was the use of layers of oil-impregnated paper tape wound round each conductor, and this type of insulation is still used for MV and HV cables. As the electrical loading on a cable varies so does the conductor temperature, and the various component parts expand and contract until eventually gaps or voids may appear in the insulation. This can lead to local electrical discharges, depending on the electrical stress, with consequent deterioration of the insulation. To overcome this problem at HV, the oil-filled cable, originally designed by Emanueli in 1926, has insulating oil fed into the centre of each conductor under pressure from oil tanks along the cable route, to fill any voids that occur.

In order to prevent water being absorbed by the insulating paper tapes the conductor/insulation assembly is covered by a metal sheath, usually lead (or lead alloy) or aluminium. Steel tape or wire armouring is applied over the lead sheath to protect it from mechanical damage, and layers of fibrous material impregnated with bitumen cover the whole assembly to prevent corrosion of the armouring or metallic sheath. Common practice is not to use armouring with aluminium-sheathed cables, but the sheath is protected against corrosion by a covering of high-quality plastic.

Originally the conducting cores of a cable were made of standard copper wire since this could be readily formed, and because copper has excellent electrical conductivity. The late 1950s saw a considerable increase in the price of copper and attention switched to the use of aluminium conductors both in stranded-wire and solid form. For equal conductance the required weight of aluminium is only half that of copper. Using aluminium conductor and sheathing results in a cable that has a smaller diameter and less weight than an equivalently rated cable with copper conductors, lead sheath and steel armouring.

Along with changes in conductor material some changes in insulation have also taken place. Thermoplastic insulants, in particular polyvinyl chloride (PVC), came into use in the 1950s. However, PVC is liable to softening at high temperature, and creates an added problem in that dense black smoke is emitted from burning PVC, as well as poisonous acidic gases. In addition, the dielectric properties of PVC are limited so that it is now only used in low-voltage cables. Polyethylene (PE) is suitable for all medium voltage levels and is solely used for insulation purposes as cross-linked polyethylene (XLPE) which allows a longer short-time heat duration under fault conditions, and a higher working temperature under normal conditions, compared with thermoplastic PE. The premature deterioration of early polymeric cables has been overcome by advances in cable design and progress in production technology.

The combination of aluminium conductors using mechanical conductor joints and XLPE insulation has been influenced by the difficulties in obtaining suitably trained staff with the skills required to join paper-insulated cables. Coupled with the use of plastic heat-shrink or rubber slip-on terminations, these developments have considerably reduced installation times and costs. XLPE is now available for use at working voltages up to 500 kV, with cross-sectional areas up to 2500 mm$^2$.

From the foregoing it will be appreciated that the construction of a cable is considerably more expensive than that of an equivalent overhead-line conductor. There is also the cost of installation, with four main methods in use. The most usual method is to lay one or more cables in a trench in the ground and refill the cable trench with a suitable backfill if the original ground is unsuitable. Other methods of installation are the laying of cables in ducts or troughs of various materials, or placing large pipes or ducts underground and pulling the cables into such pipes. The most modern method of laying MV and HV XLPE cables is by directional or guided boring. In good soil conditions some types of LV cables can be directly laid in the ground with the use of

The heat produced by the passage of current through cable conductors cannot be so easily dissipated as with an overhead-line conductor in air. With cables the insulating materials are often operating close to their maximum permitted temperature. For cables buried in soil, the rating is dependent on the maximum permissible conductor temperature, the ground temperature, the soil thermal resistivity, the depth of the buried cables, the method used to protect the cables in the trench, and the proximity of other cables carrying electrical power. Tables giving factors for deriving cable capability under a given set of conditions are available from manufacturers, cable standards and harmonisation documents, or in handbooks on cables.

Situations can occur where the limiting factor on a particular cable-conductor size is its ability to carry very heavy short-circuit currents due to a close-up fault for up to 1 or 2 s rather than maximum load current for some hours. This problem can occur when an HV/MV infeed substation is introduced at the extremity of an MV network, and for MV and LV cables close to the feeding HV/MV or MV/LV substation. While paper insulation is able to withstand a temperature of 250°C for approximately 5 s, any associated soldered conductor joints limit the maximum conductor temperature to 160°C. The use of compression cable joints raises the permissible conductor temperature to 250°C, as for EPR and XLPE cables.

In the absence of more precise technical information, eqn. 6.3 may be used to assess the maximum acceptable short-circuit current $I_{sc}$ (kA), given the conductor cross-section area $S$ (mm$^2$) and the duration of the flow of fault current $t(s)$:

$$I_{sc} = KSt^{-1/2} \times 10^{-3} \tag{6.3}$$

The coefficient $K$ depends on the initial temperature assumed, and the maximum temperature allowed for the cable under consideration. For an initial temperature of 65°C, and a final temperature of 250°C for paper and XLPE insulated paper, and 150°C for PVC insulation, the appropriate values of K are given in Table 6.4.

*Table 6.4    Values of K for determining conductor short-circuit capability*

| Conductor material | Insulation | | |
|---|---|---|---|
| | PVC | Paper | XLPE |
| Copper | 110 | 150 | 150 |
| Aluminium | 75 | 100 | 100 |

From eqn. 6.3 and Table 6.4 the maximum short-circuit current-carrying capacity of a cable with a conductor cross-sectional area of 100 mm$^2$, with protection operating in 0·6 s, is 9·7 kA for an aluminium/PVC combination and 19·4 kA for copper/XLPE.

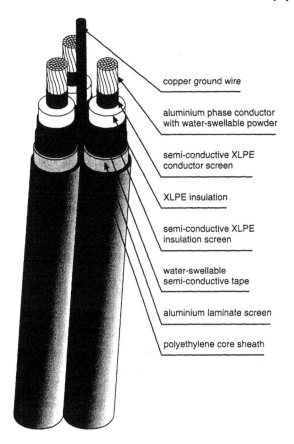

copper ground wire

aluminium phase conductor
with water-swellable powder

semi-conductive XLPE
conductor screen

XLPE insulation

semi-conductive XLPE
insulation screen

water-swellable
semi-conductive tape

aluminium laminate screen

polyethylene core sheath

*Figure 6.9    Construction of a typical MV underground cable (courtesy Nokia Cables)*

## 6.5 Switchgear

### 6.5.1 Introduction

In an electricity supply system it is necessary to disconnect equipment from the network quickly if a fault occurs in order to avoid damage to the faulted equipment or other equipment on the network. To permit equipment to be maintained it must be isolated from the live system. For operational requirements it is necessary to group circuits into specific configurations at various substations on the MV and HV networks. On the MV networks it is often necessary to vary the open and closed points to avoid excessive losses, voltage drops or fault levels or to maintain supplies under abnormal conditions. Thus various devices are required to operate an electricity supply network safely and with maximum efficiency.

*Switchgear* is a general term covering switching devices and assemblies of such devices with associated inter-connections and accessories.

A *circuit breaker* is a switching device, capable of making, carrying and breaking currents under normal circuit conditions; and also making, carrying for a specified time and breaking currents under specified abnormal circuit conditions such as those of a short circuit.

A *switch* is a switching device capable of making, carrying and breaking currents under normal circuit conditions, which may include specified operating overload conditions and also carrying for a specified time currents under specified abnormal circuit conditions such as those of a short circuit. A switch is thus, by definition, not intended to make or break fault current.

A *switch-fuse* is a switch in which one or more poles have a fuse in series in a composite unit, so that high fault currents are cleared by the operation of the fuse.

A *recloser* is a circuit breaker equipped with relays in order to carry out a variable pattern of tripping and closing.

A *disconnector* is defined as a mechanical switching device which provides, in the open position, a specified isolating distance. It should be capable of opening and closing a circuit when negligible current is broken or made. It should be noted that disconnectors are unreliable for breaking capacitative current, with capacitative currents of less than 1 A, causing damage to some types of disconnector. While it should be capable of carrying normal load current, and also carrying for a specified time currents under abnormal conditions such as for a short circuit, a disconnector is not capable of making or breaking short-circuit currents.

A *sectionaliser* is a disconnector equipped with relays so as to operate within the dead time of a recloser.

The interrelation of the operation of reclosers and sectionalisers is described in Section 7.5.

## 6.5.2 Circuit breakers

A circuit breaker has the capability of breaking fault current up to a specified value and remaking the phase connection when the fault has been cleared, using various input signals which can define the state of the circuit(s) protected by the breaker. It can repeat this operation many times between maintenance periods. In early versions the contacts were opened inside a tank of oil. When the contacts are closed within a small chamber inside the main tank a pressure is generated within the chamber by the gases formed by the arc, and this forces oil across the arc path to cause rapid extinction. In this *bulk-oil-volume circuit-breaker* arrangement the oil is used both for arc extinction and as the main insulation.

Reducing the oil content to that required for arc extinction alone resulted in the *minimum-oil circuit breaker*. Here the arc-control device is enclosed within a small housing containing oil, which is supported by an insulator to provide the necessary insulation clearance between ground and the circuit-breaker phase connections.

The use of magnetic-blowout or *air-blast circuit breakers* is rather limited in distribution systems. Most circuit breakers installed nowadays use a vacuum or sulphur hexaflouride gas ($SF_6$) as insulation, and combination vacuum/$SF_6$ breakers are also in service.

In a *vacuum circuit breaker* (VCB) the contacts open and close inside a small vacuum bottle designated as a 'vacuum interrupter'. Within the vacuum there are no ions to conduct an arc, so that contact distance for arc separation is small and there is effectively no deterioration of the contacts. The interrupter can handle high numbers of switching cycles, typically more than 10 000 operations at feeder current rating or up to 100 operations at full fault current, and experience indicates that there should be a service life well in excess of 20 years with no maintenance. Should an interrupter fail in service, replacement of the failed unit will restore the VCB back to full fault duty. The main advantages of vacuum circuit breakers are their faster operating time, lower contact wear and reduced power consumption when operating and, with fewer moving parts, little maintenance requirement. Since there are virtually no fire or explosion risks, VCBs are particularly suited to indoor installations in built-up areas.

By containing the *breaker* contacts within a compartment of $SF_6$, the gas can be forced through a nozzle at high speed into the area of the arc. The gas quickly regains its insulating qualities near current zero, thus completing the current-interrupting process. As an alternative to this puffer technique, the arc plasma can be moved by magnetic forces into a new region of fresh $SF_6$ gas. The force due to the magnetic field has an intensity related to the fault current. Thus the interrupting characteristic is controlled, within limits, by the level of current being interrupted. The insulating properties of $SF_6$ can be adversely affected by pollution and humidity, so it is essential that the gas is not contaminated. This requires particular attention to be paid to the cleanliness of maintenance equipment, and special precautions are required by maintenance staff to avoid possible contamination.

In the future, semiconductors or superconducting materials may form the basis of new methods of breaking fault current. High-current thyristors are already available and new materials are being developed which are super-conducting without having to be cooled to extreme temperatures. Conduction depends on the surrounding magnetic field. Moving from a high magnetic field to a low magnetic field would cause the superconducting material to change from a non-conducting state to a conducting state, which is the requirement for breaking fault current.

## 6.5.3 Disconnectors

Disconnectors enable off-load breaks to be made in circuits, for example when isolating sections of overhead line on rural networks. Depending on the physical arrangement, they may also provide a visual indication that an item of

circuit owing to right-of-way problems or the proximity of buildings, and lines which were originally built in fields may now be surrounded by various forms of development. Individual parts of the tower can be replaced as necessary, so that the repaired tower can then withstand the stresses of pulling in new conductors. However, re-conductoring poses further problems in that both the old and the new conductors must be kept clear of roads, railways, buildings etc., and it is very often necessary to erect considerable amounts of scaffolding or use numbers of high-level mobile platforms while re-stringing the conductors under controlled tension.

Wood poles have been used extensively at MV and almost exclusively at LV. Whilst pole lives of 80 years have been claimed, the average life of a wood pole is about 50 years. Failure is usually due to wood rot around the ground-level mark, the degree of rot being influenced by the impregnation process used and local soil conditions. Replacement is relatively easy, with the conductors being supported by temporary poles adjacent to the pole position.

Concrete structures, reinforced with steel rods, are in common use in substations. Damage can result from frost and ice, and ultimately the steel reinforcements can be exposed and suffer corrosion. The application of suitable de-rusting agents to the steel, and repair of the damaged concrete using various epoxy resin mixtures, appears to be an attractive economic solution.

When considering replacement of cables one major factor is the high costs involved. In Great Britain there are many cables 50 years or more old which were originally installed for systems operating at 5 kV, for example, uprated to 6·6 kV working, and then further uprated to 11 kV operation. There are as yet no diagnostic techniques available to predict the likely number of years of useful life remaining in these cables. A sudden increase in fault incidence is often the only indication that a cable has reached the end of its useful service life, as shown in Figure 6.11, which is taken from an actual case in Great Britain. It is generally not possible to excavate a cable and replace it by a new cable in the same trench, because of the unacceptable reduced system security during this replacement period. Where cables have been laid in ducts or pipes it may be possible to extract these and use the vacated duct of pipe for replacement cable.

Replacement of large HV/MV transformers can cost of the order of £3/4 million per unit, when account is also taken of replacing associated HV and MV connections, and relay and control equipment. While research is being carried out to equate transformer dielectric losses with deterioration of insulation, it has as yet not been possible to predict the likely future time of failure. It is unlikely that transformer replacement could be justified purely on any savings in transformer losses or reduced maintenance costs. However, when transformers of any size are replaced owing to excessive deterioration, these benefits should naturally be taken into account.

With the replacement of any equipment, the opportunity should be taken to consider whether it is economic to make provision for the possibility of uprating the network voltage level in the foreseeable future. For example, in Great Britain replacement of switchgear and cables on existing 6·6 kV networks is by 11 kV

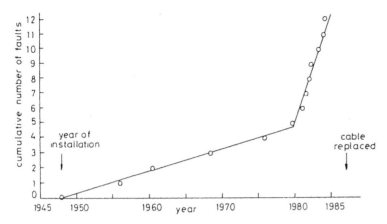

*Figure 6.11    Incidence of faults on MV cable leading to replacement*

equipment, even if the particular network is not scheduled to be uprated to 11 kV for some years. Double-ratio 11 kV/6·6 kV/LV distribution transformers can be installed in anticipation of a move to the higher voltage. Similarly, when replacing single-phase LV and MV overhead lines it would be prudent to allow for future conversion to 3-phase circuits and the possibility of uprating to a higher voltage level in the future.

Equipment replacement offers the planning and design engineer an opportunity to provide facilities for the long-term development of any system, and this aspect must always be considered rather than concentrating on a like-for-like replacement. Often it is possible to achieve a simplification of the system which improves system reliability and reduces investment expenditure. Examples of this type of development are the replacement of the double MV system, e.g. 33/11 kV in the UK, with a single voltage level, thus saving the costs of MV/MV substations by the adoption of simple satellite substations – small HV/MV or MV/LV transforming points near the load centres.

## 6.7 Bibliography

ADAM, J. F., BRADBURY, J., CHARMAN, W. R., ORAWSKI, G., and VANNER, M. J.: 'Overhead lines – some aspects of design and construction', *IEE Proc. C*, 1984, **131**, pp. 149–187
'AMKA self-supporting aerial cables', *Sähkö – Electricity in Finland*, March 1972, **45**
ASH, D. O., DEY, P., GAYLARD, B., and GIBBON, R. R.: 'Conductor systems for overhead lines: some considerations in their selection', *Proc. IEE*, 1979, **126**, pp. 333–341
ASKER, P., REGNER, L., and HEDBERG, S. E.: 'A new concept for overhead lines', *Electron. Power*, March 1986, **32**, pp. 226–228
ATKINSON, W. C., and ELLIS, F. E.: 'Electricity distribution – Asset replacement considerations', *Electron. Power*, May 1987, **33**, pp. 337–340

BALL, E. H., BUNGAY, E. W. G., and SLOMAN, L. M.: 'Polymeric cables for high voltage distribution systems in the UK'. 7th International Conference on Electricity Distribution – CIRED 1983, AIM, Liege, Paper *d*11

BERNSTEIN, B. S., and BRANCATO, E. L.: 'Ageing of equipment in electric utilities', *IEEE Trans.*, 1993, **EI–28**, pp. 866–875

BLOWER R. W.: 'Distribution switchgear' (Collins, 1986)

BLOWER R. W.: 'Experiences with medium voltage vacuum circuit breaker equipment in distribution systems', *Electron. Power*, Oct. 1981, **27**, pp. 728–731

BOSMARS, L. and SCHOONJANS, H.: 'New options for distribution transformers', IEE Conf. Publ. 197, International Conference on Electricity Distribution – CIRED 1981, pp. 15–20

'Calculation of continuous current rating of cables (100% load factor)', IEC Publication 287, 1982

CANON, D.: 'Thermal performance of underground cable', *Transm. Distrib.*, 1981, **33**, pp. 46-48

CATTARUZZA, M., MARCHI, M., and MAZZA, G.: 'Characteristics of switchgear for MV ENEL distribution network'. 7th International Conference on Electricity Distribution – CIRED 1983, AIM, Liege, Paper *e*06

'Current rating guide for distribution cables', Engineering Recommendation P/17, Amendment 1, 1978, Electricity Council, London

DEGN, W., GRASSNER, H., and WUTHRID, H. R.: 'Service experience with low oil content circuit breakers'. IEE Conf. Publ. 197, International Conference on Electricity Distribution – CIRED 1981, pp. 41–45

'Developments in distribution switchgear'. IEE Conf. Publ. 261, 1986

EVANS, J. H.: 'Replacement needs could herald new distribution designs', *Elect. Rev.*, 26 Nov. 1982, **211**, (18)

FLURSCHEIM, C. H. (Ed.): 'Power circuit breaker theory and design' (Peter Peregrinus, 1984)

FRANKLIN, A. C., and FRANKLIN, D. P.: 'The J & P transformer book', 11th edition (Butterworth, 1983)

FREWIN, K.: 'Distribution transformer development', *Electron. Power*, Feb. 1985, **31**, pp. 133-136

GARRARD, C. J. O.: 'High voltage switchgear', *Proc. IEE (IEE Reviews)*, 1976, **123**, pp. 1053–1080

GODEC, Z., and SARUNAC, R.: 'Steady-state temperature rises of ONAN/ONAF/ OFAF/ transformers', *IEE Proc. C*, 1992, **139**, (5), pp. 448–454

GULDSETH, O. J., KEGLER, W., and MULLER, K.: 'Service experience with vacuum circuit breakers'. IEE Conf. Publ. 197, International Conference on Electricity Distribution – CIRED 1981, pp. 56–60

HAZAN, E.: 'Advances in underground distribution cable', *Transm. Distrib.*, 1981, **33**, pp. 32–34

HINKKURI, A., and LEHTINEN, I.: 'SAX – a new medium voltage distribution mode'. 9th International Conference on Electricity Distribution – CIRED 1987, AIM, Liege, Paper *d* 03

HOCHART, B. (Ed.): 'Power transformer handbook' (Butterworths, 1987)

HOLMES, E. J.: 'Limits of reactance variation, tapping range and safe loading limits for large supply transformers'. IEE Conf. Publ. 197, International Conference on Electricity Distribution – CIRED 1981, pp. 305–309

KEINER, L., RAGALLER, K., and POOLE, D.: 'Service experience with and development of $SF_6$ circuit breakers employing the self extinguishing principle'. IEE Conf. Publ. 197, International Conference on Electricity Distribution – CIRED 1981, pp. 36–40

KUUSISTO, O.: 'Experiences of insulated overhead cables and new ideas in line design in the Finnish distribution system'. International Conference on Electricity Distribution – CIRED 1979, AIM, Liege, Paper 34

LANFRONCONI, G. M., METRA, P., and VECELLIO, B.: 'MV cables with extruded insulation: a comparison between XLPE and EPR', IEE Conf. Publ. 197, International Conference on Electricity Distribution – CIRED 1981, pp. 167–171

LAWSON, W. G., SIMMONS, M. A., and GALE, P. S.: 'Thermal ageing of cellulose paper insulation', *IEE Trans.*, 1977, **EI–12**

LYTHALL, R. T.,: 'The J & P switchgear book' (Butterworth, 1980)

MACKINLAY, R.: 'Managing an ageing underground high- voltage cable network', *Power Eng. J.*, 1990, **4**, (6), pp. 272–277

MCALLISTER, D.: 'Electric cables handbook' (Granada Technical Books, 1982)

MCDOUGAL, T.: '132kV $SF_6$ switchgear – an inside story'. *Power Eng. J.*, 1988, pp. 39–51

NORRIS, E. T.: 'Thermal rating of transformers', *Proc. IEE*, 1971, **118**, (11), pp. 1625–1629

O'GRADY, J.: 'Overloading power transformers – a user's viewpoint', *Power Eng. J.*, 1990, **4**, pp. 87–93

ORAWSKI, G., BRADBURY, J., and VANNER, M. J.: 'Overhead distribution lines – some reflections on design', *IEE Proc. C*, 1986, **133**, pp. 409–424

OVEREND, P. R.: 'Insulated aluminium bundled conductors for l.v. overhead mains on facades or poles', *Distrib. Dev.*, Sept. 1981, pp. 6–10

PAPADOPOULOS, M. S.: 'Polymeric insulation for 33–132kV cables in the UK', *Distrib. Dev.*, Sept. 1981, pp. 11–15

PARIS, M., LOVET, M., and MARIN, C.: 'The development of low voltage overhead systems and service connections using insulated cables', IEE Conf. Publ. 197, International Conference on Electricity Distribution – CIRED 1981, pp. 172–176

'Power cables and accessories 10kV to 180kV', IEE Conf. Publ. 270, 1986

'Power transformers: Part 1 – General', IEC Publication 76–1, 1993

PRICE, C. F., and GIBBON, R. R.: 'Statistical approach to thermal rating of overhead lines for power transmission and distribution', *IEE Proc. C*, 1983, **130**, pp. 245–256

'Revitalising transmission and distribution systems'. IEE Conf. Publ. 273, 1987

ROSS, A.: 'Cable practice in electricity-board distribution networks: 132kV and below', *Proc. IEE. (IEE Reviews)*, 1974, **121**, pp. 1307–1344

ROSS, A.: 'High voltage polymeric insulated cables', *Power Engng. J.*, 1987, **1**, pp. 51–58

ROY, M. M.: 'The electrical characteristics and selection of low-voltage power cables', *Power Engng. J.*, 1987, pp. 239–244

RYAN, H. M., and JONES, G. R.: 'SF6 switchgear', (Peter Peregrinus, 1989)

SANDERS, K. J.: 'Corrugated tanks for distribution transformers', *Electron. Power*, Oct. 1983, pp. 729–732

SAYER, W. C.: ' Application of wooden poles for distribution lines', *IEE Proc. C*, 1986, **133**, pp. 425–429

SHROFF, D. H., and STANNETT, A. W.: 'A review of paper ageing in power transformers', *IEE Proc. C*, 1985, **132**, pp. 312–319

STANNETT, A. W.: 'HV cables at high temperatures', *Electron. Power*, Sept. 1983, pp. 651–655

SWINDLER, D. L.: 'A comparison of vacuum and $SF_6$ technologies at 5–38 kV', *IEEE Trans.*, 1984, **A-20**, pp. 1355–319

'Today's transformer technology', *Int. Power Gener.*, 1991, **14**, (4), pp. 38–40

TRAVERS, R., and REIDY, P.: 'Medium voltage distribution cables using XLPE insulation. Irish utility experience'. 7th International Conference on Electricity Distribution – CIRED 1983, AIM, Liege, Paper *d*12

WATSON, W. G., WALKER, A. J., and FISHER, A. G.: 'Evaluation of the cost and reliability implications of alternative investment policies for replacement of plant on ageing distribution systems'. IEE Conf. Publ. 197, International Conference on Electricity Distribution – CIRED 1981, pp. 326–330

WEBB, D. A.: 'Ring main switchgear – expectation and solution', IEE Conf. Publ. 318, Third International Conference on Future Trends in Distribution Switchgear, 1990, pp. 12–16

WRIGHT, A., and NEWBERRY, P. G.: 'Electric fuses' (Peter Peregrinus, 1982)

ZIMMERMAN, W., and ACKERMANN, C.: 'Compact indoor switchgear installations with $SF_6$ circuit breakers for 45 to 72.5kV', *Brown Boveri Rev.*, 1981, **68**, pp. 306–310

The simplest situation is with two fuses in series, as shown in Figure 7.1a. The fuse nearest the source infeed, fuse A, is usually designated the *major* fuse, with fuse B known as the *minor* fuse. Typical operating curves are shown diagrammatically in Figure 7.1b, which indicates the required discriminating interval between the minimum melting time of the major fuse A and the maximum clearance time of fuse B. As an approximate guide successive fuses should be rated at 1·6 to 2 times the rating of their minor fuses to achieve acceptable discrimination. Fuses are in common use on LV systems and are sometimes found on MV underground networks. This practice is based on

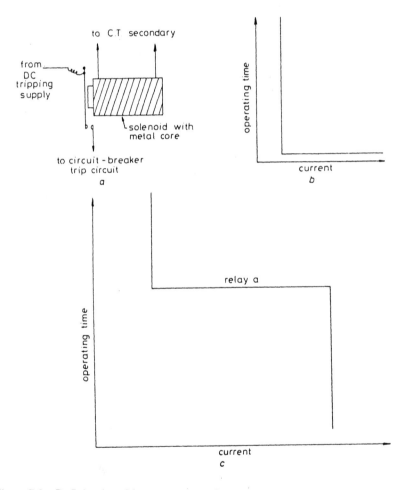

*Figure 7.2   Definite-time-delay overcurrent relays*

    *a* Relay
    *b* Time/current curve of instantaneous relay
    *c* Co-ordination of definite-time-delay relays equipped with instantaneous tripping at high currents

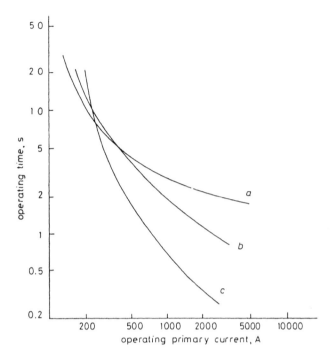

*Figure 7.3    Inverse-time-delay overcurrent-relay characteristics*

economic considerations. A more sophisticated relay protection system can be justified in MV systems since it reduces outage costs.

### 7.2.2 Overcurrent relays

The basic method of providing overcurrent relay protection on a circuit is to install current transformers (CTs) on the circuit, which then feed current into the overcurrent relay proportional to the circuit current. When the current exceeds a preset value the relay will operate at a time determined by the characteristics of the relay to initiate tripping of the associated circuit breaker.

One of the earliest forms of overcurrent relay was an instantaneous electromechanical version with a solenoid which, when sufficient current was passed through the coil from the current transformer, would magnetically attract a metal armature to operate the tripping circuit, as indicated in Figure 7.2a. This would have a time/current characteristic as in Figure 7.2b. The addition of a timing device results in a definite time-delay overcurrent relay. The time delay can be adjusted, but is independent of the value of the overcurrent required to operate the relay. By varying the time delay of successive relays in series, discrimination between the relays can be achieved. As shown in Figure 7.2c, these relays can be equipped with an additional element which provides

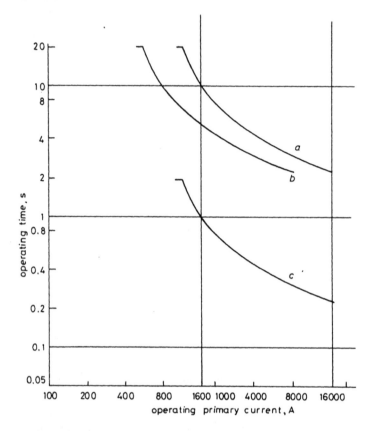

*Figure 7.4   Varying relay operating time by using different current and time settings*

   a   100% current = 800 A; 1.0 time multiplier
   b    50% current = 400 A; 1.0 time multiplier
   c   100% current = 800 A; 0.1 time multiplier

virtually instantaneous operation at a given high value of current, as indicated by the vertical line on the right-hand side of the relay characteristic. This arrangement has been used extensively in Europe, since it provides a simple solution to the problem of protection selectivity.

In the UK and USA the induction-disc relay, which has a construction similar to that of a conventional energy-consumption meter, has been used extensively. Various time/current characteristics can be achieved as shown in Figure 7.3. These relays are also designated inverse time-delay overcurrent relays, since they operate after a time delay which is inversely dependent on the value of the overcurrent. It should be noted that the time delay approaches a definite minimum value for higher values of overcurrent.

It is essential that there is adequate discrimination between relay/protection stages in series. Faults in different sections of a network will result in fault

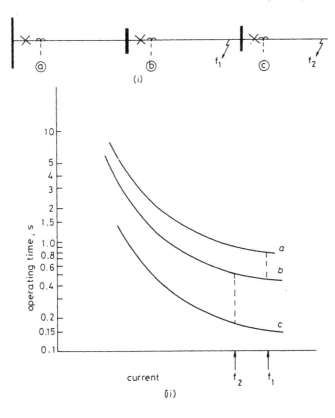

*Figure 7.5    Discrimination with inverse-time-delay overcurrent relays*

currents of different magnitudes owing to the different impedances between the power source and the fault position. If suitable relay current/time settings are selected only those circuit breakers nearest the fault will trip, leaving the other breakers to supply the healthy sections of the network. It is common practice to arrange for about 0·4–0·5s operating time between relays in series, to allow for tolerances in relay and circuit-breaker operation. With electronic relays, where the time accuracy is better, 0·3 s discrimination is possible.

For the inverse time-delay overcurrent relay the speed of operation may be altered by adjusting the sensitivity of the relay, or the time required for the relay contacts to close to trip the circuit breaker. Consider the time/current characteristic of such a relay, with a nominal operating current of 800 A primary current, and a 1·0 time multiplier, as shown in Figure 7.4a. Then, by way of example, for a fault current of 1600 A it will be seen that the relay would operate in 10 s. Altering the current sensitivity to 400 A, i.e. one-half of the previous setting, would result in the relay characteristic being moved to the left as shown by the curve *b* and the relay will operate in the same time as at setting *a* with one-half the current, i.e. 10 s at 800 A. In addition, by adjustment of the

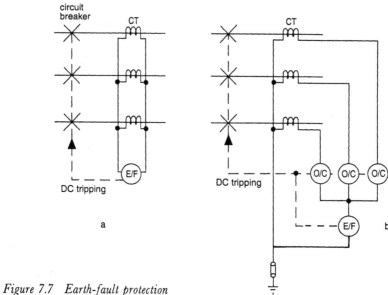

*Figure 7.7   Earth-fault protection*

E/F = earth-fault relay; O/C = overcurrent relay

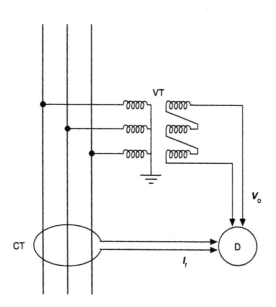

*Figure 7.8   Directional earth-fault protection: CT and VT arrangements*

D = directional element of earth-fault relay
Relay angle setting = 0° for system earthed via arc-suppression coil, 90° for unearthed system (see Figure 7.9)

With current and voltage supplied to the earth-fault relay the basic criteria for its operation are that both the asymmetric current $I_r$ and the asymmetric voltage between neutral and earth, $V_0$ have to exceed preset values. In addition, operation will only occur within a specific range of the phase angle between $I_r$ and $V_0$ as indicated diagrammatically in Figure 7.9. As discussed in Section 3.7, the fault current $I_f$ and the neutral point voltage $V_0$ both depend on the total phase–earth capacitance. The asymmetry current $I_r$ detected by the relay is a proportion of the fault current $I_f$, and is influenced by the phase–earth capacitance of the network, which will alter as the network configuration is changed.

Microprocessor-based relays can be used to apply more complex criteria, with the additional facility that any settings can be adjusted via remote-control systems. These features have extended the use of unearthed, and arc-suppression-coil-earthed, systems. On the other hand, trends to simpler and more reliable supply arrangements lead to a preference for earthed systems, where, for example, simple current sensors can replace current transformers, and where distributed automation is easier to arrange.

## 7.4 Unit and distance protection

### 7.4.1 *Unit protection*

The simpler protection facilities described in the previous sections can provide suitable discrimination for radial and small mesh networks. They do, however, have limitations in ensuring satisfactory discrimination under all the fault-level and network conditions likely to be met in interconnected HV systems, and in some exceptional combinations of circumstances. Consequently protection schemes have been devised where the protection coverage is limited to only one item of equipment.

If a fault occurs within the protected zone, relay operation should take place, whereas fault current feeding through the protected zone to a fault elsewhere on the system should not operate the protection. This form of protection is designated unit protection, since each protection arrangement protects only one unit of the system; e.g. a line or cable, a transformer, or a generator.

Unit protection schemes are based on the fact that, with a healthy circuit, the currents entering and leaving the circuit are equal. A schematic diagram of one type of this protection is given in Figure 7.10a for one phase only. Current transformers at each end of the circuit are interlinked by pilot wires as shown. The example covers the voltage-balance arrangement where the CT secondaries are connected in opposition. Current only flows through the pilot wires, and therefore through the relays, when there is a difference in the induced voltages $v_1$ and $v_2$. This will occur when there is a difference in the primary currents at each end of the circuit, as indicated in Figure 7.10b, showing a fault within the protected zone. Pilot problems limit the maximum line length which can be protected by such schemes to about 40 km, and these schemes tend to be associated with the higher-medium-voltage and high-voltage systems.

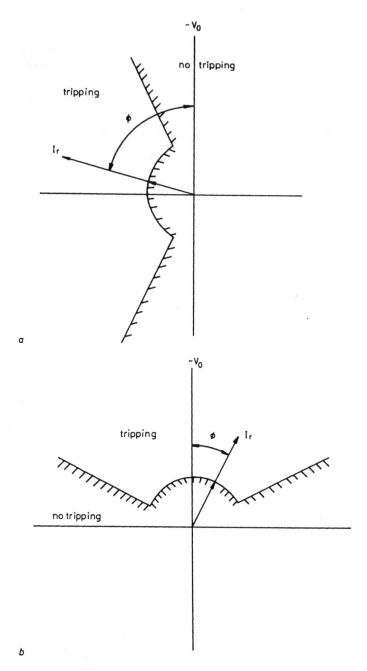

*Figure 7.9   Tripping criteria for a directional earth-fault relay*

$V_0$ = neutral-point voltage; $I_r$ = relay current
*a* Unearthed system
*b* System earthed via arc-suppression coil

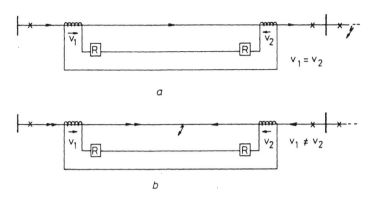

$v_1 = v_2$

*a*

$v_1 \neq v_2$

*b*

*Figure 7.10 Example of unit-protection operation*

    *a* Through fault
    *b* Internal fault, fed from two ends

Similar unit protection is used to protect transformers. The turns ratios of the current transformers have to be adjusted to take account of the main transformer voltage ratio, which can vary if the transformer has on-load tap-change facilities. In addition, the primary- and secondary-winding CT connections must take account of the main transformer winding arrangement, e.g. star/delta or delta/star.

## 7.4.2 Distance protection

The previously mentioned limitations of over-current protection, the possibility of damage to pilot-wire cables, plus possible instability on unit protection schemes with heavy through-fault currents, resulted in distance protection being developed for use mainly on HV systems. In these schemes the circuit voltage and currents are used to detect the fault, on the basis that the distance to the fault is proportional to the impedance or reactance to the fault.

Various time/distance characteristics are available for impedance protection, but stepped or sloping characteristics are most commonly used. In Figure 7.11 A, B, C and D represent the location of substations linked by feeders AB, BC and CD with a fault on feeder CD at F. Considering the stepped characteristics used in Figure 7.11, it will be noticed that a fault in the first zone covered by the protection at A, marked A1, would result in the fastest operating time of the relays at A. This zone extends for about 80% of feeder AB to avoid possible discrimination with relay B. If this 'fast zone' were extended close to substation B, variations in equipment tolerances and system arrangement could result in rapid operation of the protection at A for a fault on feeder BC close to substation B. For example, if the system is earthed at more than one point, the impedance measured by the relay system will be too low, whereas if fault arc resistance is present the fault will appear further away.

## 7.6 Overvoltage protection

As well as preventing damage to equipment due to fault current, it is also necessary to ensure that excessive voltages do not cause damage or lead to unnecessary outages. The optimum methods of protection against overvoltages, and how widely such protection should be applied, depend on system-operation practice and local weather conditions, e.g. lightning strikes. The theoretical aspects of overvoltages are discussed in Section 3.8. Some of the protection aspects will be considered in this section.

The high voltages which are induced on an overhead line from direct lightning strikes, or from strikes to nearby ground, are propagated as surges along the line and can cause damage to substation equipment unless suitable protection devices are installed. In order to prevent steep surges from entering an HV/MV substation, the HV lines are equipped with earth wires. Earth wires on MV lines close to the substations also assist. If the MV-line crossarms are earthed, this limits the overvoltage level which can be formed on the line. Often larger substations are themselves equipped with earth wires or masts to protect the substation from direct lightning strikes.

The insulation strength, or insulation level, of various items of equipment has to be greater than the magnitude of these transient overvoltages. Various overvoltage protection devices can be fitted on the network to reduce the overvoltages to an acceptable level. With the transient overvoltages limited to a given 'protection level' by such devices, the insulation level of the system must exceed the 'protection level' by a safety margin of around 20–25%.

Resonance-induced overvoltages can be prevented by avoiding exceptional operational arrangements; e.g. transformers connected to a system by their higher-, or lower-voltage windings only. A broken conductor can lead to resonance overvoltages. Since the operation of a phase fuse creates a similar situation, omitting fuses on MV systems has the advantage of limiting such overvoltages. The jump resonance of voltage transformers can be dampened to a tolerable level by adding a resistor to the secondary winding, or open tertiary delta, of the voltage transformer (VT). There is no effective protection against sustained overvoltage, so that it is vital to design any system so that excessive overvoltages of this type cannot be produced.

The simplest form of overvoltage protection is the so-called air gap with one metallic rod connected to a phase conductor and a second rod connected to earth. The gaps are frequently connected across an insulator, as, for example, on the higher-voltage side of a transformer. The setting of the air gap is made such that flashover occurs at 70% or less of the rated impulse-withstand voltage. Any such flashover then becomes an earth fault on the system and is usually cleared by an auto-recloser operation. Air gaps are inexpensive, but do not fully protect against very steep-fronted surges induced by direct or indirect lightning strikes.

This limitation has led to the use of lightning arresters, which have the further advantage that their use avoids the follow-on power current experienced with air gaps. A conventional arrester consists of a bushing with a number of air gaps and non-linear resistors in series between the phase and earth connections. The resistance of the resistor decreases very rapidly as the current through it increases, so that, when the gaps break down on overvoltage, the resistors ensure that the voltage on the conductor is limited to a safe value. Once the surge has been discharged, the voltage across the arrester starts to drop and the current falls. The resistance of the resistor then increases and the follow-through power current is sufficiently reduced to be broken by the air gaps at the next voltage zero. Operation is fast and the unit is then once again available for overvoltage protection duty. The voltage and current characteristics of lightning arrester operation are shown in Figure 7.13, where the sharp fall in current following breakdown of the gap can be seen.

The arresters with series air gaps have two shortcomings. The voltage stress on the protected equipment rises to a very high value before sparkover occurs, and the sparkover then causes a sudden drop in voltage, as seen in Figure 7.13. To minimise sudden changes in voltage stresses, a gapless type has been developed, based on the characteristics of non-linear resistors, and is in common use. The modern metal-oxide arrester, without any gaps, leads to a smoother and more effective method of protection.

Lightning arresters are most effective when located as close as possible to the equipment they are protecting – no more than 10 m away. Longer distances lead to higher voltages being imposed on the equipment. They are primarily used at

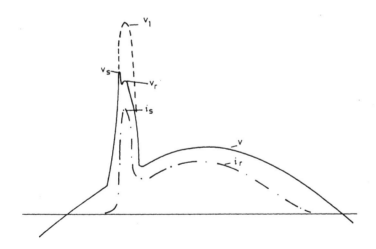

*Figure 7.13    Voltage/current characteristic of a lightning arrester*

$v$ = normal phase-earth voltage waveform
$v_1$ = voltage surge if no lightning arrester used
$v_s$ = lightning-arrester breakdown voltage
$v_r$ = residual voltage
$i_s$ = peak current
$i_r$ = follow-through current

both sides of HV/MV substations and on cable/overhead-line junctions. Overhead-line/cable surge impedances are approximately in the ratio 10:1.

In some areas where the keraunic level is high, as in central USA for example, each MV/LV distribution transformer may be fitted with a lightning arrester. While this may not save the transformers close to a lightning strike on the MV circuit, the arresters prevent damage to many units from overvoltages on the network caused by direct or indirect strikes. In countries with lower keraunic levels it is often economic to protect the larger distribution transformers connected to overhead lines with arresters, while protecting the smaller units with air gaps.

## 7.7 Automation of network operation

The main purpose of automation is to obtain better system performance and to improve the reliability of supplies to customers, by faster clearance of faults and restoration of supplies.

There are many levels of automation. Two simple examples of automatic devices are fuses and air gaps which operate when specified current and voltage levels are exceeded. Protection relays and lightning arresters are more advanced versions of these devices. Various decentralised automatic arrangements have been in service on distribution systems for some decades. One of the earliest was the use of automatic tap changers and voltage regulators for voltage-drop compensation. With conventional distribution network arrangements any faulted MV feeder was identified solely from the operation of the protection relays, and accurate location of the fault was not easy. The introduction of devices on the networks which indicate the passage of fault current, known as fault indicators, has assisted considerably in fault location and isolation of the faulted section.

The use of microprocessor-based relays, which can measure a number of input signals to derive the required operating sequence for the specific fault condition, as well as having in-built self-checking facilities, has resulted in sophisticated protection and fault-clearing schemes being developed. Increasing use of microprocessor logic-controlled sectionalisers is removing the dependence on utility control staff intervention, leading to more rapid isolation of faults and restoration of supplies. Telecontrolled disconnectors, distributed around the MV network, are a further extension of automating system operation to reduce down time for fault clearance or network maintenance, or optimising network flows to reduce system losses, for example. Using suitable computer hardware and programs, network configurations can be automatically re-arranged on the occurrence of faults to minimise the consequences of further system outages.

The advent of telecommunication channels to individual substations made it possible to provide more instruction codes to more equipment, and to receive information back on the state of the equipment. Thus, a single relatively low-powered transmitter operating at a frequency of a few hundred megahertz can provide communication channels between local control centres and individual substations, and also between a central control point and several local control centres. With two-way links each substation is equipped with its own receiver plus a small power transmitter to send back data to the control centre.

With the two-way facility available, system control and data-acquisition (SCADA) systems can be set up, where not only can instructions be transmitted to specific items of equipment at every telecontrol substation but also many and varied types of data can be transmitted from each substation to the control centre. Circuit-overload and fault data, including equipment and protection faults, are instantly transmitted to the control centre, triggering an alarm to alert the control engineer to a fault on the electrical network or one of the auxiliary support systems. The necessary action can then be taken to isolate faulty equipment and restore the network to a satisfactory configuration. Computer-based SCADA systems make it possible to pre-program various system-control operations to minimise down time in the even of a fault, or to optimise network security and/or system losses while at the same time capturing real-time system data for interrogation and storage. Apart from sending back real-time information on the state of each network, such systems can collect energy and demand data from points on each network. Linked to a computer, such information can be processed to provide useful data for statistical analysis, as discussed in Chapter 11.

## 7.8 Distribution management systems

Arising from the developments referred to earlier, one of the most interesting areas has been the development of *distribution management systems* (DMS) utilising the traditional SCADA system, real-time fault-current information from relays, and the *network information system* to provide support tools for control centre operators in order to minimise the total operational costs (power losses and outage costs) subject to certain technical constraints being kept within acceptable values.

Figure 7.14 shows a model for defining the various states of a distribution network. The normal state can be divided into two parts. The unacceptable section excludes those disturbed states detected and alarmed by SCADA functions but includes those constraint violations which must be detected by network analysis, while the acceptable state is divided into optimal and non-optimal states. Figure 7.14 also shows the close relationship between planning and operation. The functions in the normal state are generally planning tasks while those outside it relate more to operational problems.

In the disturbed state the automation system detects an abnormal network event due to a fault, or some system limitation being exceeded, and sends an alarm. The loss of a feeder would be signalled as a faulted state but a feeder which has been faulted or switched open for maintenance purposes, and is now

option using heuristics. The user can choose one option and the system analyses the effect of the control and displays the results. The heuristic optimisation is based on real-time and forecast network calculations and can be interactive or independent.

The interactive planning of a maintenance outage begins when the user lists the circuits or busbars to be de-energised. The system analyses the target state and proposes the best alternative based on heuristics, but here also the user can choose another one. The selected switching sequence is then carried out on the network model and network calculations are executed using the estimated loading on the network for specified periods. If there are any problems, the system presents the next options and proposes the best one but ultimately the operator is responsible for opening or closing remote controlled disconnectors or advising staff to operate the manual disconnectors.

When a distribution system transfers from the normal state, the SCADA system obtains information from the process. In the case of a permanent fault, the primary system transfers to the faulted state and some operations such as event analysis, fault location, fault isolation and restoration are needed. The objective regarding operations in the faulted state is to minimise the outage costs of customers. In this analysis the faulted feeder, the type of fault, and the fault current measured by the fault indicators are determined. The analysis can also produce an assumption of the cause of the fault.

The fault location module uses the results of event analysis, network data and the heuristic knowledge of operators to suggest possible fault locations. First the measured short-circuit current is compared with the short-circuit calculations, and the data on the operations of fault detectors are used. The possible locations are classified using terrain and weather data, and heuristic rules. In the module fuzzy reasoning is used to model any inexact knowledge.

Confirmation of the fault location is obtained by interactive experimental switching measures to isolate the faulted line sections. The system gives relevant information on the outage areas and back-up connections and proposes the best option, minimising the outage costs, but again the user can choose any other one. The impact of the chosen operations on the network is analysed and the constraints are checked. The interactive process continues until a successful arrangement is obtained. In both the fault isolation and restoration procedure operations are carried out in two phases, with remote controlled operations being performed before manual operations. In addition the road network in the area is taken into account for examining various routes to each manually controlled disconnector and using the fastest one in order to assess the optimal restoration time.

The use of the intelligent application functions of the support system increases the ability to minimise the total costs subject to technical constraints. The use of the functions in the faulted state reduces the outage costs while the optimisation function reduces losses. Also constraint violations are efficiently detected with this system.

# 7.9 Bibliography

ANDERSON, R. B.: 'Lightning performance criteria for electric power systems', *IEE Proc. C*, 1985, **132**, pp. 298–305

'Application and coordination of reclosers, sectionalisers and fuses'. IEEE PES Tutorial Course 80 EH0157-8-PWR, 1980

'Application guide for the selection of fuselinks for high voltage fuses for transformer circuit applications', IEC Publication 787, 1983

AUBREY, D. R.: 'The ECRC 11 kV automatic sectionaliser for the protection of overhead lines', *Distrib. Dev.*, June 1982, pp. 22–27

BAIGENT, D., and LEBENHAFT, E.: 'Microprocessor-based protection relays: design and application examples', *IEEE Trans.*, 1993, **IA–29**, pp. 66–71

BAUMGARTNER, H.: 'Report on an automatic earth fault indicator for high-voltage systems operated with the neutral insulated'. IEE Conf. Publ. 250, 8th International Conference on Electricity Distribution – CIRED 1985, pp. 154–159

BERGEN, L.: 'Satellite control of electric power distribution', *IEEE Spectr.*, 1981, **18**, (6), pp. 43–47

BERGGREN, J.: 'Highly sensitive earth-fault protection using a dedicated microprocessor'. 7th International Conference on Electricity Distribution – CIRED 1983, AIM, Liege, Paper c09

BICKFORD, J. P., and HEATON, A. G.: 'Transient overvoltages on power systems', *IEE Proc. C*, 1986, **133**, pp. 201–225

BJÖRNSTAD, T., and PAULSEN, H. M.: 'Control methods for the distribution overhead line network'. 10th International Conference on Electricity Distribution – CIRED 1989, pp. 540–544

BLACKBURN, J. L.: 'Symmetrical components for power systems engineering' (Marcel Dekker, USA., 1993)

BRADWELL, A.: 'Electrical insulation' (Peter Peregrinus, 1983)

BURKE, J. J.: 'Philosophies of overcurrent protection'. Proceedings of joint international power conference, Athens Power Tech – APT'93, 1993, pp. 304–308

CASSELL, W. R.: 'Distribution management systems: functions and payback', *IEEE Trans.*, 1993, **PWRS–8**, (3), pp. 796–801

CASTELLAZZI, G., RETUS, A., and TONON, R.: 'Automatic fault location and isolation in medium voltage power distribution networks'. IEE Conf. Publ. 250, 8th International Conference on Electricity Distribution – CIRED 1985, pp. 286–291

CEGRELL, T.: 'Power system control – experiences and future perspective'. Proceedings of joint international power conference, Athens Power Tech – APT'93, pp. 660–664

CIESLEWICZ, K., and MARCINIAK, L.: 'Central automatic location of short-circuits in branched MV cable distribution systems'. 7th International Conference on Electricity Distribution – CIRED 1983, AIM, Liege, Paper *e*16

CLARKE, G. J., and ROSEN, P.: 'The ring connecting unit with full range fuses – a new concept in system protection'. 9th International Conference on Electricity Distribution – CIRED 1987, AIM, Liege, Paper *e*16

CRUCIUS, M., and REICHELT, K.: 'Surge diverters and over voltage limiters in MV-systems'. IEE Conf. Publ. 197, International Conference on Electricity Distribution – CIRED 1981, pp. 114–118

DARVENIZA, M.: 'Lightning protection of power systems', *J. Elect. Electron. Eng. Aust.*, 1991, **11**, (3), pp. 201–209

DIALYNAS, E. N., and MACHIAS, A. V.: 'Interactive modelling of substation operations following a failure event', *IEE Proc. C*, 1987, **134**, pp. 153–161

ERIKSSON, A. J., and MEAL, D. V.: 'The incidence of direct lightning strikes to structures and overhead lines, IEE Conf. Publ. 236, 1984, 'Lightning and power systems', pp. 67–71

FEENAN, J.: 'The versatility of high-voltage fuses in the protection of distribution systems', *Power Eng. J.*, 1987, **1**, pp. 109–115

FUJII, Y., MIURA, A., HATA, Y., TSUKAMOTO, J., YOUSSEF, M. G., and NOGUCHI, Y.: 'On-line expert system for power distribution system control', *Elect. Power Energy Syst.*, 1992, **14**, (1), pp. 45–53

GARDINER, M. J.: 'The protection of the overhead system in Northern Electric', *Power Eng. J.*, 1992, **6**, (1), pp. 9–16

GOSDEN, J. H.: 'Lightning and distribution systems – the nature of the problem', IEE Conf. Publ. 108, 1974, Lightning and the distribution system, pp. 1–8

HEADLEY, A., BURDIS, E. P., and KELSEY, T.: 'Application of protective devices to radial overhead line networks', *IEE Proc. C*, 1986, **133**, pp. 437–440

'High voltage fuses: Part 1 – current limiting fuses', IEC Publication 282–1, 1985

'Report on h.v. fuse links for the protection of ground mounted transformers'. ACE Report 86, 1983, Electricity Council, London

HUENCH, F. J., and WRIGHT, G. A.: 'Fault indicators: types, strengths and applications', *IEEE Trans.*, 1984, **PAS–103**

HUSE, J.: 'Lightning overvoltage and overvoltage protection of distribution transformers', IEE Conf. Publ. 197, International Conference on Electricity Distribution – CIRED 1981, pp. 110–113

JÄRVENTAUSTA, P., VERHO, P., and PARTANEN, J.: 'Using fuzzy sets to model the uncertainty in the fault location of distribution feeders', *IEEE Trans.*, 1994, **PWRD–9**, (2), pp. 954–960

KARTTUNEN, M., and LAINE, P.: 'Eliminating ferroresonance oscillation in a voltage transformer', *Sähkö – Electricity in Finland*, 1970, (3)

KATO, K., NAGASAKA, H., OKIMOTO, A., KUNIEDA, T., and NAKAMURA, T.: 'Distribution automation systems for high quality power supply', *IEEE Trans.*, 1994, **PWRD–6**, (2), pp. 954–960

KELSEY, T., and BESFORD, S. J.: 'The use of modern reclosers and sectionalisers in rural distribution networks', *Electron. Power*, Oct. 1983, **29**, pp. 724–726

KUAN, K. K., and WARWICK, K.: 'Power distribution network design aided by an expert system', *Power Eng. J.*, 1990, **4**, pp. 79–86

LAYCOCK, W. J.: 'Management of protection', *Power Eng. J.*, 1991, **5**, (5), pp. 201–207

LEMOINE, J. C., MESSAGER, P., and SACHER, Y.: 'The French policy concerning automatic switching and remote control devices on the supply network'. International Conference on Electricity Distribution – CIRED 1979, AIM, Liege, Paper 55

'Lightning and the distribution system'. IEE Conf. Publ. 108, 1974

'Lightning and power systems'. IEE Conf. Publ. 236, 1984

MARLOW, D. E., and DAUNCEY, R. G.: 'An approach to sequence switching using computerised telecontrol', IEE Conf. Publ. 266, 1986, Power system monitoring and control, pp. 52–56

McMILLAN, R., FORMBY, J. R., and WILSON, R. G.: 'Restructuring the distribution network by the use of feeder automation', IEE Conf. Publ. 273, 1987, 'Revitalising transmission and distribution systems'. pp. 146–150

MORAN, R. J., DECESARO, F. P., and DUGAN, R. C.: 'Electronic sectionalizer control methodology for improved distribution system reliability', *IEEE Trans.*, 1992, **PWRD–7**, pp. 876–882

OAKES, M.: 'Improved protection in h.v. distribution systems', *Elect. Rev.*, 1986, **218**, pp. 14–15

'Power system protection: Parts 1, 2 and 3' (Peter Peregrinus, 1981)

'Protective relays application guide', GEC Measurements (Baldini and Mansell, 1982)

REDFORD, S. J.: 'The rationale for demand-side management', *Power Eng. J.*, 1994, **8**, (5), pp. 211–217

ROSS, D. W., CHOEN, A. I., PATTON, J., and CARSON, M.: 'New methods of evaluating distribution automation and control (DAC) systems benefits', *IEEE Trans.*, 1981, **PAS–100**, pp. 2978–2987

ROSEN, P.: 'Distribution system protection – the way ahead', *Electron. Power*, May 1987, **33**, pp. 333–336

RUDOLF, D.: 'The use of fibre optics at electric utilities, as illustrated by HEAG'. 9th International Conference on Electricity Distribution – CIRED 1987, AIM, Liege, Paper d14

SIDHU, T. S., SABHDEV, M. S., and WOOD, H. C.: 'Microprocessor based relay for protecting power transformers', *IEE Proc. C*, 1990, **137**, pp. 436–444

SMITH, H. L.: 'DA/DSM directions. An overview of distribution automation and demand-side management with implications of future trends', *IEEE Comput. Appl. Power*, 1994, **7**, (4), pp. 23–25

STEWART, J. S.: 'An autorecloser with microprocessor control for overhead line distribution', *Electron. Power*, June 1984, 30, pp. 469–472

'Surge protection in power systems', IEEE Tutorial Course 79 EH0144–6–PWR, 1978

TEO, C. Y., and CHAN, T. W.: 'Development of computer aided assessment for distribution protection', *Power Eng. J.*, 1990, **4**, pp. 21–27

VAN DE WATER, C. J., and VAN OIRSOUW, P. M.: 'Probabilistic overload assessment in distribution networks'. IEE Conf. Publ. 338, Third International Conference on Probabilistic Methods Applied to Electric Power Systems – PMAPD 91, 1991, pp. 163–168

VAN SOM, P. J. M., and CESPEDES, R. H.: 'Improved tools for real-time control', *Transm. Distrib. Int.*, Second Quarter 194, **4**, pp. 42–49

VERHO, P., JÄRVENTAUSTA, P., KÄRENLAMPI, M., and PARTANEN, J.: 'Intelligent support system for distribution management'. International Conference on Intelligent System Application to Power Systems, Montpellier, France, 1994, pp. 699–706

WHITE, T. M.: 'Method of protecting distribution transformers connected solidly to an urban 11kV network', *IEE Proc. C*, 1986, **133**, pp. 441–444

WINTER, K.: 'Swedish distribution networks – aspects on neutral treatment, earth fault clearance and related matters'. IEE Conf. Publ. 318, 1990, Third International Conference on Future Trends in Distribution Switchgear, pp. 127–129

WRIGHT, A.: 'Application of fuses to power-system equipment', *Power Eng. J.*, 1991, **5**, (3), pp. 129–134

YOUNG, D. J., LO, K. L., McDONALD, J. R., and RYE, J.: 'Development of a practical expert system for alarm processing', *IEE Proc. C*, 1992, **139**, (5), pp. 437–447

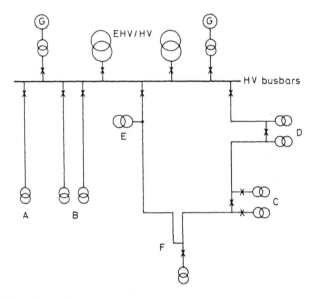

*Figure 8.2  Alternative feeding arrangements to HV/MV substations*

    X  circuit breaker
    ⊖  transformer

switch and a circuit breaker with current transformers. A bypass facility has been provided to permit servicing of the 110 kV circuit breaker without disconnecting the transformer. When the circuit breaker is bypassed a fault-making switch is used for back-up protection in case a fault occurs while the circuit breaker is being serviced.

HV circuits are usually provided on overhead lines except in urban areas or where permission for overhead lines cannot be obtained; e.g. in some areas of outstanding natural beauty or where the number of overhead lines around an EHV/HV substation is considered excessive. Various types of overhead-line construction have been referred to in Section 6.3. The construction adopted in a particular locality will depend to a considerable extent on the materials available locally, the above-mentioned amenity requirements, and the local ground and weather conditions. For long overhead lines the three phases of an HV line may be transposed at regular intervals in order to achieve symmetrical phase impedances.

In city areas with high load densities the optimal size of HV/MV substations is higher than in rural areas. If the HV/MV voltage ratio is not too high then MV circuits may provide a useful means of providing back-up capacity to the HV supply. In many countries the division of responsibility is often at some point on the HV system, which may cause organisational problems when considering the overall simultaneous planning requirements for the HV and MV systems.

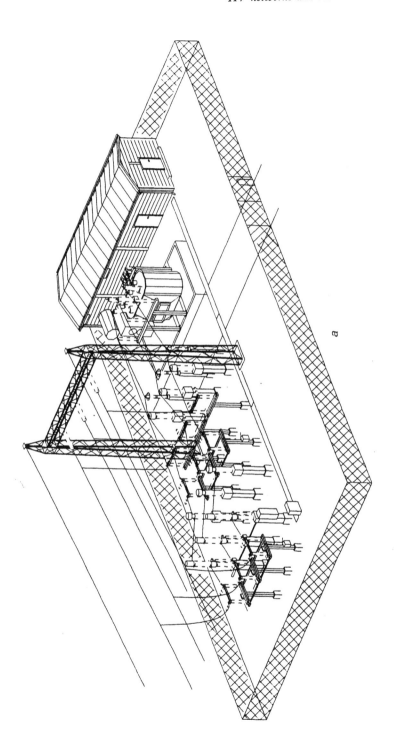

*a*

*Figure 8.3a   Layout of a simple one-transformer HV/MV substation (Courtesy ABB Finland)*

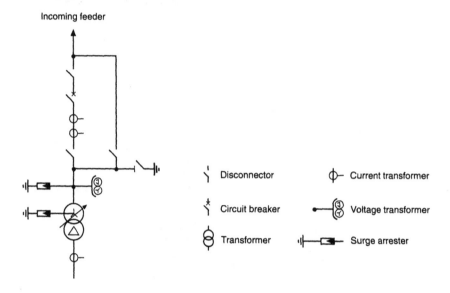

Incoming feeder

| | | |
|---|---|---|
| ) Disconnector | ϕ⊢ Current transformer | |
| ) Circuit breaker | ⊸⊗ Voltage transformer | |
| ⊗ Transformer | ⊣⊏▪ Surge arrester | |

*Figure 8.3b    Simple one-transformer HV/MV substation (Courtesy Strömberg Co.)*

## 8.3 Substation arrangements

### 8.3.1 Site location

The importance of HV/MV substations to the quality and security of supply has already been mentioned. It follows that the siting and timing of these substations is one of the most essential aspects of power-system planning, especially when the overall cost of such a project is taken into account. It is therefore vital that all the options being considered must be capable of coping with the long-term provision of electricity supplies in the area; i.e. upwards of 25 years in the future. This involves consideration of the initial and ultimate requirements for HV feeders, HV/MV transformers and substation switchgear arrangements, as well as adequate provision for the associated MV equipment. All the options considered must be assessed from both technical and economic viewpoints.

The precise location of a new HV/MV substation is influenced by many factors. It should always be placed as near as possible to the centre of gravity of the load, or the expected centre of a high-load-density area. Attempting to supply too many load centres from one substation often turns out to be false economy. Solving such problems involves long-term studies, and sensitivity analysis applied to load growths and equipment costs.

The difficulty of obtaining land at the exact required location can make it impossible to achieve the optimum site technically and economically, and this is another aspect to be included in any sensitivity analysis of the economics of supplying the area. With a limited number of sites of the right size to choose from, the choice must take account of the total cost of developing the site; i.e. the cost of bringing in new circuits plus the civil work and building costs, including earthing costs. Poor site conditions, resulting in increased civil-engineering costs, may be offset by reduced line costs, but it is the total construction cost for each site, including the line work, that is important in the overall economic assessment.

If the site is some distance from the existing HV network, considerable costs can be involved in extending circuits to the new substation. Equally, the cost of extending numbers of MV circuits will exclude certain potential sites. The optimum location, economically, is thus determined by the number of HV and MV circuits involved, and is influenced by being a rural or urban situation, since this then affects the relative overhead/underground costs, as discussed in Section 6.3.

Having established the technical and economic consequences of factors such as those mentioned above, it is then possible to set out 'balance sheets' for the various sites being considered, having particular regard to the forecast long-term future of the substation, in order to determine the optimal site. An example of optimising substation location is described in Section 14.5.2.

## 8.3.2 Design philosophy

Substations are convenient points for the control and protection of the transmission and distribution networks. In their design a number of interrelated problems have to be solved before a layout can be achieved which is both technically sound and economically attractive. Wherever possible, at the high-voltage level it is common practice to install equipment outdoors provided that there are no major environmental or weather constraints.

The substation equipment includes circuit breakers to interrupt fault current, disconnectors to isolate circuits for maintenance, plus busbars and other connections to interconnect the circuits, as well as supporting structures, insulators and various auxiliary systems, and generally power transformers also. All these items have to be co-ordinated to provide a suitable and satisfactory substation arrangement. It must also be remembered that every item of equipment added to a substation arrangement can itself introduce further possibilities of failure which must be considered against the overall substation operational performance.

A large number of different electrical arrangements are available to provide maximum facilities at minimum cost. In all cases consideration must be given to such factors as whether the substation can be easily extended in the future. The substation should be capable of easy maintenance without danger to staff or any interruption of supplies, and there should be alternative facilities available in the case of an outage, whether caused by a fault or required for maintenance purposes.

## 8.3.3 Mesh-type substations

The earlier UK HV systems utilised mesh substations to minimise the amount of switchgear required at any one site. In Figure 8.4a a 3-switch open-mesh

substation is shown located on a feeder between two HV substations. The dashed lines indicate the equipment within the mesh substation, while only the relevant section of the substations at each end of the feeder has been shown. This 3-switch open-mesh arrangement was normally used to control two circuits and two local HV/MV transformers. With this arrangement a fault on a transformer can be cleared by tripping the transformer HV circuit breaker, leaving the HV circuit arrangements unaffected. A circuit fault is cleared by tripping open the bus-section circuit breaker BS and the associated transformer HV circuit breaker, as well a the far-end circuit breaker as shown in Figure 8.4*b*. Continuity of supplies at the substation is retained by the remaining circuit and transformer.

With this arrangement, maintenance of the bus-section circuit breaker would involve interrupting the HV through circuit. A bypass arrangement is therefore sometimes installed to alleviate this situation. By closing the bypass disconnectors and isolating the HV bus-section circuit breaker, as shown in Figure 8.4*c*, continuity of the through circuit is retained during the maintenance outage of the bus-section circuit breaker. Where an additional HV/MV transformer is required to provide security of supplies on loss of one of the loaded transformers, but cannot be permanently switched through to the MV switchboard owing to the resultant excessive fault levels on the MV switchgear, a third transformer can be connected to the bypass busbar but isolated from the HV mesh as shown in Figure 8.4*d*. On the outage of one of the transformers, owing to fault or maintenance, the standby transformer can be put into service by re-arranging the isolation facilities as shown in Figure 8.4*e*.

An extension of the 3-switch arrangement already discussed is the open-mesh arrangement shown in Figure 8.4*f*, where provision is made to control three circuits, plus three local transformers each having individual HV circuit breakers, so that any fault on a transformer does not break the continuity of the HV network connections. Figure 8.4*g* illustrates the 4-switch closed-mesh arrangement, capable of controlling four circuits plus four local transformers or transformer feeders. As Figure 8.4*h* shows, these mesh arrangements can get quite complex, and one major problem with mesh-substation arrangements is the difficulty in extending the substation if provision was not included in the original layout.

The advent of HV automatic motorised disconnectors has made it possible to reduce switchgear at mesh substations, with arrangements such as that shown in Figure 8.4*i*. For a fault on a circuit or transformer, the far-end circuit breaker and the substation bus-section circuit breaker open to isolate the fault. Under 'dead' conditions the appropriate disconnector is automatically opened to isolate the faulty equipment from the rest of the system and the circuit breaker(s) reclosed to restore the unfaulted circuit or transformer back to service.

### 8.3.4 Single- and double-busbar arrangements

The single-busbar arrangement shown in Figure 8.5*a* is the simplest configuration to provide a convenient method of operation. Here a number of incoming

MV feeders are bussed together with local HV/MV transformers. With this arrangement a circuit or transformer has to be taken out of service to enable maintenance of the associated circuit breaker to be carried out. In addition, any extension of the busbar would require a complete shutdown of the substation. Furthermore a busbar fault would cause all circuit breakers to trip, isolating the switchboard. By adding a bus-section circuit breaker as in Figure 8.5*b*, a busbar fault or work on the switchboard to extend a busbar only leads to the loss of one-half of the circuits connected to the substation. Grouping incoming and outgoing circuits evenly across the two sections of busbar, and ensuring that where feeders go to the same load point they are connected to separate sections of the busbar, further improves security. Further improvements can be achieved by wrapping around the ends of the single busbar to obtain the ring busbar shown in Figure 8.5*c*. Here additional busbar disconnectors have been provided to ensure adequate electrical clearance when maintaining the busbar disconnectors, and then only one circuit has to be taken out of service during maintenance.

By providing an auxiliary transfer busbar and a bus-coupler circuit breaker BC to the original single-busbar layout, as in Figure 8.5*d*, a circuit can be selected to the auxiliary busbar by suitable use of the busbar-selector disconnectors, e.g. circuit 1. The circuit is then protected by the bus-coupler circuit breaker BC. Circuit breaker CB can then be isolated as shown for routine

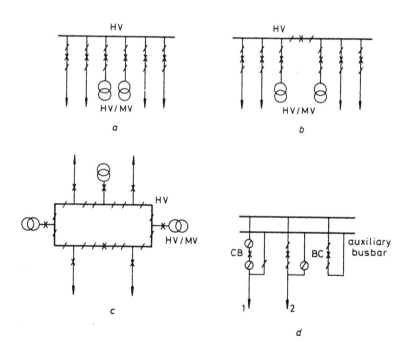

*Figure 8.5    Single-busbar arrangements*

maintenance or repair, or examination after fault clearance with the circuit still in operation.

The use of two busbars with isolating facilities, such that each circuit can be selected to either busbar, is shown in Figure 8.6*a*. It is then possible to arrange the various circuits into different configurations on the two busbars, as required for operational reasons. It is also possible to select all circuits to one busbar, releasing the other busbar for maintenance such as insulator cleaning.

The addition of a bus-coupler circuit breaker, as shown on the right-hand side of Figure 8.6*b*, enables on-load transfer of a circuit from one busbar to the other to be carried out. The installation of a bus-section circuit breaker in one of the busbars provides improved circuit marshalling facilities, and permits segregation of circuits to minimise the effect of busbar faults.

For reasons of space, 'wrap round' arrangements are often adopted with the reserve busbar encircling a central main busbar. Complex configurations are possible. By way of example, Figure 8.6*c* shows 12 circuits connected to an arrangement with a total of six sections of busbar. Only busbar disconnectors are included for clarity. It will be noted that busbar sections A1 and A3 are connected together, with sections A2, B2 and B3 separately joined, while section B1 is isolated from all other sections. Thus circuits 1, 3 and 8 are grouped together, circuits 7, 2, 9, 5, 12, 11 and 10 are grouped together separately, with only circuits 4 and 6 on the isolated busbar B1.

The above example is typical of the variations that can be achieved in system configurations using double-busbar layouts, and which make it possible to cater for a wide variety of system operational requirements. It is often usual to connect two circuits from one area to different sections of busbar, in order to avoid loss of supply on busbar faults. The number of incoming lines, generators and transformers connected to a particular section of busbar, or group of busbars, can be restricted to avoid excessive phase–earth or 3-phase fault currents. Sections of the network can be isolated from the rest of the system; e.g. industrial plant causing fault-level or nuisance load problems. Generators can be selected to particular circuits to force power into a given area. This facility to rearrange circuit configurations also assists in carrying out maintenance work on busbars without interruption of supplies.

The introduction of compact metal-clad $SF_6$ switchgear has made it possible to install large HV substations in urban areas either within buildings or underground to avoid environmental problems. The reduction in space occupied by 145 kV gas-insulated switchgear (GIS), compared with that for a conventional open-terminal arrangement, is shown in Figure 8.7. The respective bay widths are 1·4 m and 10·5 m. A cross-section of a phase-integrated 145 kV GIS bay is shown in Figure 8.8. Typically this would be 3·3 m high, with a bay width of 1·4 m and depth of 6 m.

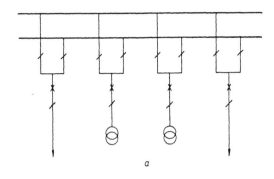

*a*

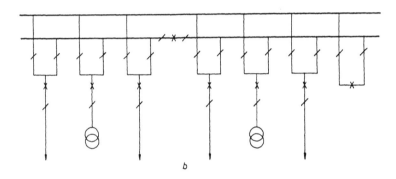

*b*

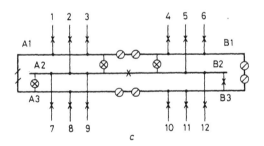

*c*

*Figure 8.6    Double-busbar arrangements*

X  circuit breaker (closed)
⊗ circuit breaker (open)
∕  disconnector (closed)
⊘ disconnector (open)
⁀  automatic motorised disconnector
⊖ HV/MV transformer

information on particular aspects of HV substation equipment not covered here, the reader is referred to the Bibliographies associated with Chapters 6 and 8.

### 8.4.2 Transformer protection

Reference was made to transformer unit protection in general terms in Section 7.4.1. Figure 8.10 refers to a differential protection scheme across a star–delta transformer. To compensate for the primary phase change it will be noted that the protection current transformers are delta-connected on the primary side, and star-connected on the secondary side. To avoid the possibility of relay operation owing to transformer magnetising current inrush on switching in the transformer, restraining coils sensitive to the third-harmonic components of current are generally included. The current transformers are often located in the main-transformer primary and secondary bushings to provide overall protection. More modern relaying systems include the facility to programme the protection system to compensate for the various transformer winding arrangements whilst retaining the same CT configurations on both windings.

Whilst differential protection can detect severe internal faults, on occasion a fault between a few turns of a winding can occur. Eventually these faults cause the transformer insulation to decompose and release gas. Appreciable quantities of gas may be produced before the fault is detected by the differential protection

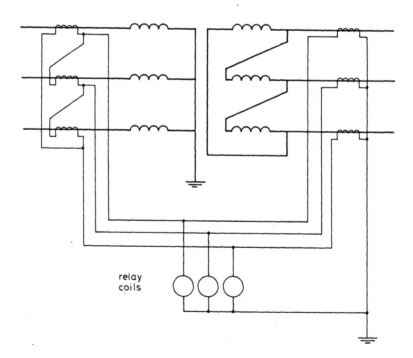

relay
coils

*Figure 8.10   Transformer differential protection*

relays. A Buchholz relay connected in the oil pipework between the top of the tank and the oil conservator traps any gas. Eventually sufficient gas accumulates in the Buchholz relay chamber so that the level of oil falls, causing a float alarm relay to operate. A heavy fault which causes a surge of oil or gas to flow through the Buchholz relay operates a trip relay to open both the HV and MV circuit breakers.

As mentioned in Chapter 6, large transformers have a thermal time constant of some hours, so that short-time overloads should not trip out a transformer.

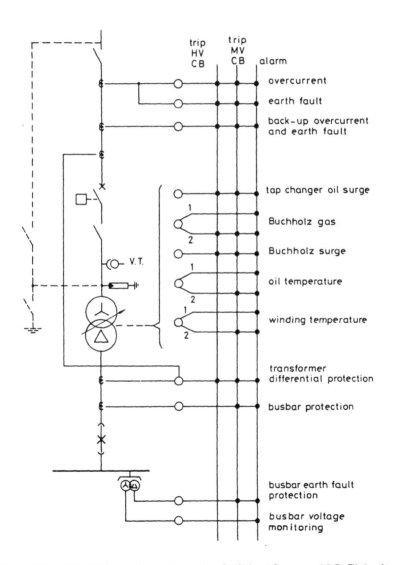

*Figure 8.11    HV/MV transformer protection facilities (Courtesy ABB Finland)*

However, a relatively slow cumulative effect of excessive load may slowly overheat the windings and insulation. Therefore a suitably dimensioned resistance element, which closely models the response of the transformer winding which has a time constant of between 5 and 10 min, is supplied with current proportional to the load current from a separate CT. Thus the temperature-sensing element is influenced by both the oil and load temperatures. Normal practice is for this relay to set off an alarm at a temperature where load can be reduced without damaging the transformer, say 85°C. Should the temperature rise higher, say to 100°C, the lower-voltage circuit breaker is tripped open to remove the heat-producing load.

The tap changer is also vulnerable to faults. However, being located within the transformer differential-protection zone, any electrical fault is detected by this protection. Figure 8.11 shows schematically the protection facilities usually installed to cover various transformer faults.

### 8.4.3 Busbar protection

Fault levels at substation busbars are usually higher than elsewhere on the system. Consequently faults at substations can often cause serious damage to equipment, be a source of potential danger to staff and buildings, and seriously affect the reliability of supply. To protect against busbar faults a form of differential protection is used where incoming and outgoing feeder flows are summed. A busbar fault would result in an imbalance in the summed CT currents, causing the protection to operate to ensure that all circuit breakers connected to the faulty busbar are opened to clear the fault. Because of the importance of not isolating busbars unnecessarily, it is usual to provide two sets of busbar protection. Tripping of circuit breakers is then only initiated when both sets of protection operate.

In addition, mechanical, electrical and electromechanical interlocking systems are in use to prevent incorrect interconnection of equipment; e.g. to ensure that disconnectors cannot be opened on load, or energised circuits be connected to earth.

### 8.4.4 Auxiliary supplies

Low-voltage power is required at substations to supply various auxiliary services. Low-voltage AC is used for the lighting, heating, cooling and ventilation of buildings, together with supplies to auxiliary equipment such as transformer cooler fans and pumps, and tap changer motors, and the charging or DC batteries. Usually the LV AC supplies are obtained from MV/LV auxiliary transformers associated with the HV/MV transformers, with alternative supplies available from the utility's LV network adjacent to the substation, wherever possible. Typically the AC system is 400/230 V of 50–200 kVA capacity, depending on the size of the substation and the amount of equipment.

DC low-voltage supplies are required to provide power for the operating mechanisms of circuit breakers, for operating alarm, protection and control systems, and for emergency back-up lighting. Typical DC voltage levels used at substations are 48, 110 and 220 V. The battery and the associated charger capacities are determined from the average load in association with the peak demand. Duplicate DC systems are sometimes installed at the larger substations to ensure high reliability.

## 8.5 Bibliography

DOWELL, M. W.: 'Compact HV substation for urban power distribution', *Elect. Rev.*, 1979, **224**, pp. 23–25

DREYFUS, H. B.: 'Bulk supplies to conurbations in Great Britain'. CIGRE Paper 32–02, 1974

DE HAAS, J., BIELDERS, M. J. J., SIDLER, H. F. A., VOS, C. W. M., and ZANTIGE, J. F.: 'Comparative analysis of different types of substation configurations'. CIGRE Paper 23–110, 1984

KLING, W. L.: 'Interaction between transmission and distribution planning'. Report n. 37–94, (WG.07) 30E, Final report of CIGRÉ/CIRED WG CC.01, Paris, August 1994

NATIVEL, J. M., and BORNARD, P.: 'New designs of HV/MV substations'. 9th International Conference on Electricity Distribution – CIRED 1987, AIM, Liege, Paper *e*11

*Table 9.1    Relative characteristics of medium-voltage overhead lines.*

| Voltage, kV | Feeder length, % | Power losses per km, % | Capital cost per km, % |
|---|---|---|---|
| 6·6 | 41 | 278 | 93 |
| 11 | 100 | 100 | 100 |
| 20 | 331 | 30 | 112 |
| 33 | 900 | 11 | 131 |

Table 9.1 shows the advantage of obtaining a considerably increased capability for a small increase in capital costs which are themselves offset by reduced power losses. By taking into account the various cost components and the voltage-drop constraint, comparisons can be made of the cost of providing a feeder plus associated distribution transformers and the relevant portion of the main infeed substation costs for various feeder operating voltages.

Figures 9.1 and 9.2 are based on studies carried out to assess the optimum medium-voltage level in a number of countries where the existing small networks needed to be extensively developed. In deriving these diagrams, in addition to the costs of the MV lines, the costs of the associated HV/MV and MV/LV substations have been included. Figure 9.1 indicates that, based on a 20 km line, 33 kV is cheaper than 20 kV for line loadings above 100 kW/km. Figure 9.2 shows that the profitability of a single MV level 33 kV system, compared with an 11 kV system, is sensitive to the load density, the average length of feeders and the estimated annual load growth. For example, at an estimated annual load growth of 15%, 20 km feeders at 33 kV would be economic for load densities above 30 kW/km, while at a 3% per annum growth the limit would be 90 kW/km. The former limit would, however, lead to excessive voltage drops as indicated by the broken lines which show the effect of a 10% voltage drop so that only about 10 kW/km could thus be accepted for an 11 kV line.

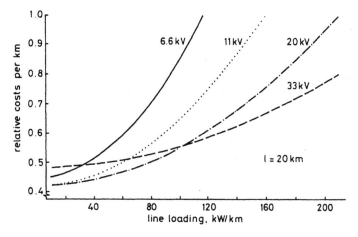

*Figure 9.1    Relative costs for 20 km overhead lines*

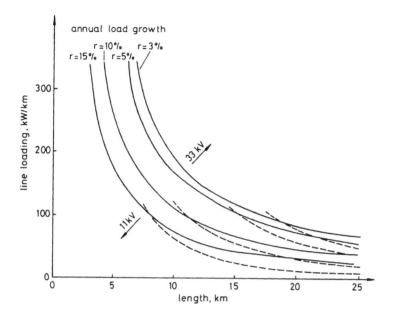

*Figure 9.2   Feasibility studies for 11 kV and 33 kV feeders*

11 kV 95mm$^2$ Al; 33 kV 35mm$^2$
— — limit based on voltage drop
——— limit based on economic considerations

   The decision of the voltage level to be selected must be based on long-term studies. When considering the electrification of developing countries, it is necessary to consider the potential loads in areas which will not be covered in the initial stages of the electrification programme, so that the long-term requirements of each area are properly catered for. In existing networks it is generally not possible to economically justify a voltage uprating in the short term (5–10 years) because of the high investment required initially to set up the higher-voltage network, especially in built-up areas. It is seldom reasonable for an individual utility to select a voltage level which is not being used by other companies in the same country, or to consider the use of non-standard voltages. The use of a common voltage level should lead to savings in component prices owing to the large volumes being produced, plus increased competition between manufacturers.

## 9.3  System modelling

In addition to the correct choice of the voltage level for an MV system, other technical policies have to be agreed. Examples of these include the use of underground cables or overhead lines, neutral earthing practice and protection arrangements, and the eventual future use of any existing single-phase spurs. Any policies on the valuation of losses or non-distributed energy, or variations in local economic conditions or safety regulations, have an effect on planning philosophies, and many of these decisions are interrelated. In many utilities most of these policies will probably be already fixed, especially in industrialised countries with a long-standing electricity supply.

Considering the fixed design policies, it is possible to construct simplified techno-econometric models for the MV system. These models can include simplified mathematical descriptions of such items as:

- Area load models
- Network configurations
- Any technical constraints
- Cost models for components, and for supply interruptions

Typical results of this type of study would include the following items, as functions of various parameters such as load density:

- Optimum sequence of voltages
- Optimum density of HV/MV substations
- Optimum sizes of transformers
- Optimum standard sizes of cables and lines

It is also often possible to obtain the sensitivities of such items to variations in economic parameters and technical constraints.

Although the mathematical equations used in such studies are generally as simple as possible the results are often complex, owing to the large number of factors involved, so that the results are usually presented graphically. However, it is sometimes possible to develop very approximate formulas. For example, the following formulas may be used for an urban area for estimating the optimum density and sizes of substations:

$$S_s \simeq k_1 \sigma^{1/3} \quad \text{and} \quad L/S_F \simeq k_2 \sigma^{-1/2}$$

where       $S_s$ = optimum ratings of HV/MV substations, MVA

$k_1, k_2$ = coefficients depending on local cost relations and MV arrangements ($k_1 \simeq 50$, $k_2 \simeq 3$)

$\sigma$ = area load density, MVA/km$^2$

$L$ = optimum length of MV feeders, km

$S_F$ = feeder load, MVA

Instead of using simple network configurations and very approximate load models, the system model can be based on typical existing systems. Also, in such a case simple cost functions can be used for substations and circuits, but average circuit lengths and the location of concentrated loads would be obtained from actual systems, preferably by using information collected from network-monitoring computer programs and regression analysis.

Results from these model-based studies can be used to determine the most appropriate electricity-supply arrangements for a developing country. These studies are also useful when considering various long-term scenarios for a particular area. The model can provide an indication of the required number of substations, and the total costs associated with different development projects.

Alternative methods are also available to provide such information. These are based on case-oriented computer-aided simulations, as discussed in Chapter 14. For most MV planning tasks, CAD systems provide more reliable results than those obtained from the above-mentioned approximate formulas or system modelling techniques.

## 9.4 Principles for determining circuit dimensions

Determining the dimensions of the conductor cross-sectional area for a new spur or a section to be replaced is a commonplace task. In this section some fairly simple rules are introduced. A more difficult but important task is the consideration of a single circuit section as a link in the whole network; e.g. the influence it has on the voltage drop and other characteristics of the entire route, under both normal and back-up conditions.

In determining the dimensions of a new section it is necessary to meet the requirements of the overall general plan for the network. For example, the future provision of a new primary substation will limit the use of the smaller sizes of cables or lines around the proposed location owing to the increase in fault level when the substation is connected into the network. This is one of a number of constraints which has to be considered. The most economic size of conductor must be chosen from those sizes which are acceptable from all points of view. In an overhead line, if the conductor has been dimensioned on an economic basis, there will usually be a large margin between the full-load rating and the thermal limit. However, extremes of ambient temperature need to be carefully considered.

The *voltage drop* is an important constraint in designing MV overhead networks. There is no agreed exact value of the voltage drops that can be accepted, either in the overall distribution system or allocated to the individual voltage levels. Often a specific voltage drop, say 5%, is permitted for the whole MV network. Thus, depending on the estimated load growth and required life of the system, the maximum voltage drop for present loading conditions will be somewhat lower. A detailed consideration of network voltage theory and practice will be found in Chapter 13.

The voltage drop $V_d$ can be calculated using the following formula from Section 3.4:

$$V_d \simeq \sqrt{3}I_p R + I_q X) \simeq \frac{P}{V}(R + X \tan \phi) \qquad . \qquad (9.1)$$

It will be noted that, in Table 9.2 and Figure 9.3, the type and size of the conductor is represented by a code. ACSR 54/9 represents a conductor with 54 mm$^2$ of aluminium conductor wire strands plus 9 mm$^2$ of steel core reinforcement, while AAC 132 refers to a conductor with 132 mm$^2$ of aluminium conductor wire strands.

*Table 9.2   Conductor data*

| Conductor | Thermal limit MVA | $R$, $\Omega$/km | Investment costs, £/km |
|---|---|---|---|
| ACSR 34/6 | 7·4 | 0·848 | 7870 |
| ACSR 54/9 | 9·9 | 0·536 | 9200 |
| AAC 132 | 17·5 | 0·219 | 11470 |

By way of example, if the first year demand $S_1$ is 1.6 MVA and the estimated annual load growth is 4%, the AAC 132 conductor is the most economic, but if the load growth is only 2% per annum ACSR 54/9 should be selected. Comparing Figure 9.3 with Table 9.2 shows the considerable difference between the thermal limit and the economic loads. Similar studies and dimensioning rules can also be carried out for underground or overhead cables, LV lines and different types and sizes of transformers.

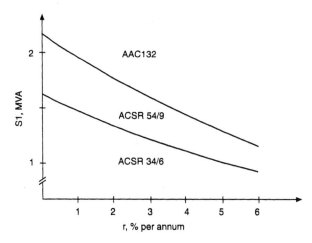

*Figure 9.3   Dimensioning guide for conductors for new 20 kV line*

$S_1$ = first-year demand
$r$ = annual load growth, %

## 9.4.2 Example: Choice of conductor size to replace an existing conductor

In this example it is required to determine the optimum demand for conductor replacement. One justification for replacing a conductor by a larger-size one is to reduce the cost of the losses by more than the investment costs of the conductor replacement. Figure 9.4 illustrates annual savings in losses if a conductor is replaced by a larger one. The larger the line loading the greater will be the savings in losses, since they are proportional to current squared. The annuity of the investment is also shown in the Figure. The optimum demand for conductor replacement is when the annual savings in losses are equal to the annuity of the investment, i.e. when

$$\frac{S^2}{V^2}\Delta R L c_l = \varepsilon c_I L \tag{9.6}$$

where  $c_I$  = investment cost of replacing conductor $a$ by conductor $b$, £/km

$\varepsilon$  = annuity factor = $(p/100)/[1 - \{1/(1+p/100)^t\}]$

$\Delta R = R_a - R_b$

$R_a$  = resistance of conductor $a$, $\Omega$/km

$R_b$  = resistance of conductor $b$, $\Omega$/km

From eqn. 9.6 the optimum demand for conductor replacement is given by

$$S = V\sqrt{\frac{\varepsilon c_I}{c_l \Delta R}} \tag{9.7}$$

The optimal load for the conductor replacements, given in Table 9.3, can be determined, when using the data from the example in Section 9.4.2 and the investment costs for conductor replacement given in Table 9.3. The investment costs for conductor replacement include material and installation costs of a new conductor and costs of pulling down an old conductor. The residual value of an old conductor has been also taken into account when defining the investment costs for conductor replacement. The resistance of ACSR 21/4 is 1·360 $\Omega$/km.

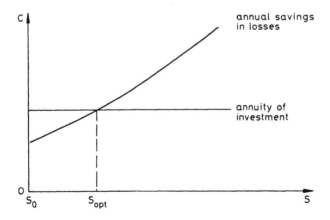

*Figure 9.4   Annual costs which have to be compared for line reconductoring studies*

Table 9.3 shows, for example, that, if the load exceeds 1·8 MVA, replacement of ACSR 21/4 conductor could be either by ACSR 54/9 or by AAC 132 conductor. Both alternatives are economically equal. In this example, for ACSR 34/6 conductor, the most suitable replacement at all loads above 2·4 MVA is AAC 132. If the reason for replacement is not purely economic, say because the existing conductor size is inadequate for the prevailing fault current, it is possible that the smaller replacement conductor may be more economic.

*Table 9.3    Optimum load levels for replacement of 20 kV overhead-line conductors*

| Conductor change | | Investment costs, £/km | Load level MVA |
|---|---|---|---|
| Existing | New | | |
| ACSR 21/4 | ACSR 54/9 | 4140 | 1·8 |
| ACSR 21/4 | AAC 132 | 5870 | 1·8 |
| ACSR 34/6 | ACSR 54/9 | 4060 | 2·8 |
| ACSR 34/6 | AAC 132 | 5790 | 2·4 |
| ACSR 54/9 | AAC 132 | 5650 | 3·3 |

## 9.5 Operational aspects

### 9.5.1 Network arrangements

When planning the arrangement of each MV system, the general targets for satisfactory system performance at minimum cost, given in Chapter 2, should be kept in mind. Details of the various calculations are given in Chapters 4 and 5 as well as in Sections 9.3 and 14.6; so the requirements for optimum conditions will be discussed qualitatively rather than quantitatively here.

The majority of the outage time that customers experience is due to faults occurring on the MV system. Various sectionalising and load-transfer facilities are used to limit the effects of MV faults. Operating arrangements for overhead-line and underground-cable networks differ considerably. The economic and thermal limits of individual cables used in city centres are close, and this has an influence on the size of cables used for back-up purposes. However, in other urban areas, the relationship is greatly influenced by the cost of laying cables, the load density and the rate of return adopted for the economic appraisal. Overhead systems are more prone to transient faults, which has led to the development of rapid autoreclosing schemes for these systems.

Almost without exception MV networks are operated radially, although an interconnected method of operation could lead to lower voltage drops and losses. However, in radial networks the protection arrangements are simpler, voltage control is easier and fault levels are lower, and if construction costs alone were considered, the preferred network arrangement would be radial.

## Overhead networks

The most common network configuration is the partially looped network with interconnections to other networks fed by neighbouring primary substations. Under normal conditions, by opening switches or disconnectors at the appropriate points, the network can be operated as a radial arrangement with a high degree of reliability. The effect of a fault on a looped circuit or interconnector can then be limited to customers on the section of the network which is taken out of service by opening disconnectors or switches to clear the fault.

Figure 9.5 shows schematically typical arrangements for a rural overhead radial feeder, with some of the manually operated disconnectors omitted for simplicity. It will be noted that each main trunk feeder has a number of lateral spurs. Except in the more remote areas supplied by such a system, there would usually be the facility of interconnection to other MV feeders, supplied either from the same, or adjacent, HV/MV substations, as indicated in Figure 9.5.

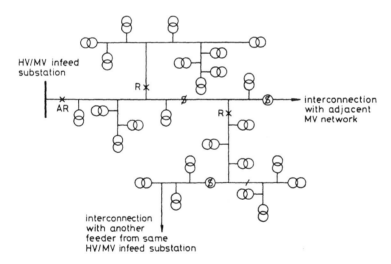

*Figure 9.5 Schematic of typical MV overhead radial feeder*

| | |
|---|---|
| X | circuit breaker |
| AR | autoreclosing facilities fitted |
| R | pole-mounted autorecloser |
| / | disconnector, normally closed |
| ⊘ | disconnector, normally open |
| ⌀ | automatic sectionaliser |
| ⊖ | pole-mounted transformer |

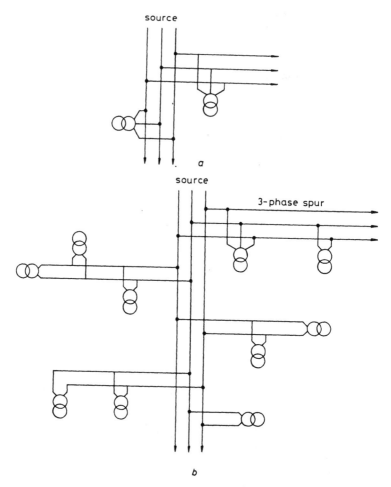

*Figure 9.7    MV 3-phase and single-phase systems*

Single-phase systems may provide a cheaper method of supplying rural areas with a low load density. Where it is likely that the network loads may increase significantly in the future, the overhead line should be constructed so that conversion to a 3-wire 3-phase system can be achieved at minimum additional cost when network loadings or voltage-drop considerations justify the uprating. It should be noted that, with the heavy demands now taken by many LV domestic customers, a single-phase LV supply is often inadequate, so that a 3-phase LV supply may be necessary. However, with a single-phase MV supply it is not possible to provide this 3-phase LV supply from a single-phase MV system without the expense of uprating the MV and MV/LV supplies to the customer from single-phase to 3-phase.

*Underground networks*

In urban areas with more heavily loaded MV systems, the main streets and roads may have one or more MV feeder cables buried below the pavements. In very heavily loaded areas it may be necessary to lay cables below roadways, owing to lack of space under the pavements because of the number of other services (gas, water, sewers, telephone, low-voltage cables etc.) already installed there, and/or the narrow width of the pavements.

In the open-loop configuration shown in Figure 9.8*a*, the various MV/LV distribution substations which are supplying individual customers or general distribution load have some form of disconnecting device on the incoming and outgoing feeder and on the transformer itself. For example, substations C and D are shown equipped with ring-main units, described in Section 10.2. At a suitable point on the network the loop is opened by a sectionalising device S, which may be a circuit breaker, switch, fuse or link. The system then effectively operates as two radial feeders.

On the occurrence of a fault, e.g. at F, the source circuit breaker A would open to remove the fault from the system and supplies would be lost to all customers between A and S. In the example shown, when the faulty section has been located, it can be isolated from the system by manually opening the isolating devices *c* and *d*. Closing S would restore supplies back up to D. Reclosure of A restores supplies between A and C, so that, under the arrangement shown, supplies would eventually be restored to all customers leaving the faulted section of cable isolated for repair work.

Developments in 3-phase $SF_6$ load-break switches, which can be accommodated in small underground chambers, permit auto-sectionalising to be used. In Figure 9.8*b*, the distribution substations contain a feeder sectionalising unit (FSU). With the introduction of microprocessor-controlled switch actuator units and remote fault-section indication (RFSI) equipment, both the in-line disconnector and the transformer disconnector can be automatically controlled by various signalling arrangements, in a manner similar to that already in use for automatic sectionalisers on overhead systems.

On the occurrence of a fault, the source breaker (1) would open. With the section of the MV feeder now de-energised, the RFSI (2), having registered the passage of fault current, would cause the in-line disconnector in the FSU on the upstream side of the fault at substation B (3) to open. The transformer disconnector at substation B would then be transferred from the downstream MV feeder to the upstream feeder (4), and the source circuit breaker can then be closed to restore supplies to the substations A and B. At substation C the transformer disconnector (5) would also be transferred to the downstream MV feeder. The MV in-line feeder isolator downstream of the fault (6) would then be opened under de-energised conditions to isolate the fault. With no voltage being registered on the MV section between the isolated faulted section and the open point (7), the open point would be closed to restore supplies to substation C from the adjacent feeder.

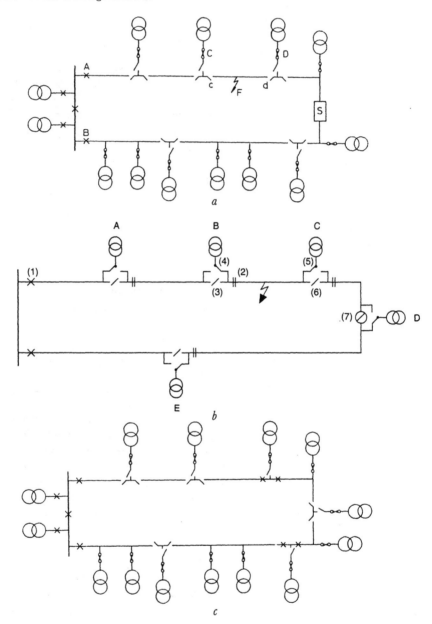

*Figure 9.8   MV underground loop arrangements*

X   circuit breaker
⌐⌐   load-breaking switch
o∞o   fuse
⊕   customer or distribution transformer
S   isolation point
−/−   disconnector
−⫤−   fault current detector

By introducing circuit breakers into a loop network, the totally closed-loop arrangement shown in Figure 9.8*c* is obtained. The ratio of MV/LV distribution substations with circuit breakers, to those without circuit breakers, will generally be a utility policy decision. The inclusion of circuit breakers reduces the downtime of the non-faulted section of the network for any fault on the loop, with sections of the feeder being automatically isolated by opening the circuit breakers on either side of the fault. Such arrangements also aid fault location, and reduce the amount of manual switching required to transfer load off the faulted MV section.

## 9.5.2 Load-transfer schemes

Where an individual HV/MV transformer supplies an isolated section of MV busbar, a fault on the transformer, or on an HV circuit solely supplying that transformer, can lead to a large number of customers being off supply for a long period of time. To reduce the effects of such faults, a number of arrangements have been devised to transfer the supply from the faulted transformer circuit to the remaining transformer unit(s) and/or alternative infeed circuits. These alternative feeding arrangements are normally capable of meeting the total demand, although on occasions there may be some limitation in the back-up supply, generally due to larger voltage drops than normal being experienced.

By way of example, consider the arrangement shown in Figure 9.9*a*, where the two HV/MV transformers are operated independently on the MV side, e.g. to avoid excessively high MV fault levels. On the loss of transformer T1 or T2, the MV bus-section circuit breaker 3 is automatically closed, either via auxiliary contacts on circuit breakers 1 or 2 as appropriate, or from loss of voltage signal from a VT on the MV side of T1 or T2. Closure of circuit breaker 3 restores supplies.

As discussed in Section 12.2.3, arc-furnace and similar interference loads often need to be segregated from the normal distribution loads to avoid affecting the quality of supply to the normal distribution customers. It is then sometimes necessary to provide duplicate transformers for both the distribution load and for the interference load to ensure independent security of supplies to both loads; i.e. a total of four transformers. However, consideration may be given to the arrangement of Figure 9.9*b*, where the distribution and interference loads are supplied by three transformers. Under normal operating arrangements two transformers would be switched to supply the largest load, say the distribution load, in order to minimise series losses. This is achieved by operating with circuit breaker 3b open.

Should transformer T2 be faulted, for example, circuit breaker 2 would trip open and this action would cause circuit breaker 3a to open and 3b to be closed, restoring supplies within a few seconds. This arrangement requires only three transformers, instead of the four-transformer arrangement referred to in the previous paragraph. Since the loads are carried by only three transformers in this arrangement, the total series losses will be greater than with the

adequate quantitative reliability analysis, as discussed in Chapter 4. The optimum types and locations of equipment can be considered either by carrying out individual cost/benefit analysis or by using optimisation methods.

In an individual analysis, costs of outages on the network are first evaluated before any new equipment is added, and the studies are then repeated with any proposed equipment installed at various alternative locations. These alternatives are then assessed by comparing the differences in overall costs. This method is adequate if not too much new equipment is to be added. Such a situation often arises when existing networks are being considered. The various types of new equipment to be used, and their locations, can be selected heuristically by the engineer, or they can be based on the results of a reliability analysis. These cost/benefit analyses can be carried out either manually (see example in Chapter 4), or more effectively by using an adequate reliability calculation program as described in Section 14.5.2.

When new networks are being designed, or large amounts of new equipment are being added, it is often beneficial to use methods whereby the amount and locations for the new equipment can be evaluated more simply. It is then possible to use approximate methods based on making some simplifications about the networks. Approximations that can be made for mathematical modelling and optimising are, for example, that the demand, and the failure rate per unit of line length, are constant. When the amount of equipment required has been approximately calculated, the planning engineer can place the individual items at suitable locations on the network. He can then refine the study by justifying the cost/benefit of each item, and optimising its location as illustrated in the following example.

*(i) Optimum location of a disconnector*

The optimum location of a pole-mounted disconnector is determined by comparing its benefit, in terms of the reduction of outage costs, at various possible locations. The basis of the method is that the new equipment protects customers upstream of its location, i.e. nearer the source, from faults downstream. The location is thus determined by the product of the transformer kVA capacity upstream, and the length of line (km) downstream.

Without a switch in the feeder circuit, a fault anywhere on the feeder will cause an interruption of supply to all customers supplied from that feeder equal to the repair time. After addition of a disconnector, then any faults downstream of its location will result in loss of supply to the customers upstream only for the length of time taken to reclose the disconnector (switching time). If we assume that:

- Risk of failure is proportional to length of line
- Costs of outages are proportional to energy not supplied
- Growth of load is equal at every load point
- Average values are used

the savings in the total costs can be expressed as:

$$C_i = KP_i f_l l_i t c_n - C_a \qquad (9.8)$$

where $C_i$ = savings in total costs if a new disconnector is at location $i$

$\quad\quad K$ = coefficient by which the first-year outage costs are to be multiplied in order to obtain discounted costs over the whole review period

$\quad\quad P_i$ = demand upstream of $i$

$\quad\quad f_l$ = failure rate per line length

$\quad\quad l_i$ = line length downstream of $i$

$\quad\quad t$ = the difference between the repair time and the switching time

$\quad\quad c_n$ = outage-cost parameter of energy not supplied

$\quad\quad C_a$ = costs of a disconnector (including maintenance) over the whole review period

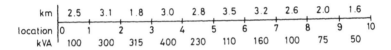

Figure 9.10 *Example of determining the optimum location of a disconnector*

As an example consider the line shown schematically in Figure 9.10. The possible positions for a disconnector are at locations 1–9. The total nameplate ratings of the distribution transformers and the line lengths for each section are also shown in the figure. The data to be used for solving eqn. 9.8 for this example are:

$K = 10$
average demands $= 0.25$ times the nameplate ratings
$f_l = 0.125$ faults/km per annum
$t = 2h$
$c_n = £0.5/\text{kWh}$
$C_a = £3750.$

The values for $C_i$ for the different locations are given in Table 9.4. It will be seen that the best site for a disconnector is at location 4 with total savings of £1720. However, it will often be necessary to check if the addition of two or more disconnectors, instead of only one, would give still better savings. In this example, the optimum arrangement is for the one disconnector only.

This type of study is sensitive to the values applied to the parameters of eqn. 9.8, as shown in Figure 9.11, where the value $c_n$ of energy not supplied has been varied. For a value twice that used in the above calculations for Table 9.4, it seems likely that more disconnectors could be justified, whereas at half the value there is no justification for placing a disconnector at any of the specified locations.

*Table 9.4   Determination of optimum location on a disconnector*

| Location | Upstream, kVA | Downstream length, km | kVA km | $C_i$, £ |
|---|---|---|---|---|
| 1 | 100 | 23·6 | 2360 | −3010 |
| 2 | 400 | 20·5 | 8200 | −1190 |
| 3 | 715 | 18·7 | 13370 | 430 |
| 4 | 1115 | 15·7 | 17505 | 1720 |
| 5 | 1345 | 12·9 | 17350 | 1670 |
| 6 | 1455 | 9·4 | 13677 | 520 |
| 7 | 1615 | 6·2 | 10013 | −620 |
| 8 | 1715 | 3·6 | 6174 | −1820 |
| 9 | 1790 | 1·6 | 2864 | −2850 |

The optimum location of other equipment, e.g. telecontrolled disconnectors, and circuit breakers, can be calculated in a similar way to those used in the above two examples. In each case it is necessary to take account of the appropriate annual costs, the duration of outages and outage rates, and possibly also system losses, in order to derive the total costs.

In addition to the location and type of new equipment to be installed, the timing of such work should also be optimised. If it can be assumed that the savings achieved with new equipment will not decrease in the future, the optimum installation time is when the annual savings are equal to, or greater than, the annual costs of the new equipment.

### 9.5.4 Network reinforcements

MV network reinforcement is motivated by three different factors. The necessity to replace equipment reaching its technical or economical life has been discussed in Chapter 6. Reinforcement due to changes in load level has been referred to earlier in this chapter. In this section the need to reinforce because of a revision of the MV system will be considered.

A voltage-system revision usually involves simplification of the existing system or the introduction of a new higher-voltage level. These are easier to carry out on overhead systems, but in any case the necessary work can take many years and requires large capital investments. Figure 9.12 gives an example of the time required for such an operation, covering a situation where a 110/30/10 kV system was replaced by a 110/20 kV system. The revision proved highly successful, giving payback times as short as five years in many rural areas.

Where large-scale equipment changes or extensions are necessary it is often economically viable to uprate, e.g. 6·6 kV networks to 11 kV operation, owing to the saving in system losses even before the additional capability at the higher voltage is required. By good planning it is often possible to install switchgear, cables and dual-ratio distribution transformers on the system some years before the voltage uprating is to take place, in order to ease both the financial and

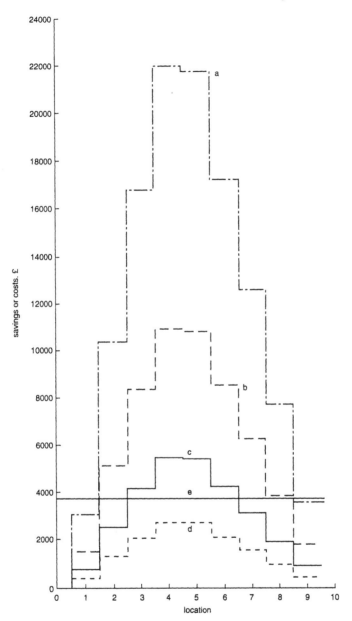

*Figure 9.11*    *Savings, or costs, gained by installation of a disconnector*
Location nodes refer to Figure 9.10
Savings in costs of outages if
*a*  $c_n = £2/\text{kWh}$
*b*  $c_n = £1/\text{kWh}$
*c*  $c_n = £0.5/\text{kWh}$
*d*  $c_n = £0.25/\text{kWh}$
*e*  Costs of a disconnector

# Chapter 10

# Distribution substations and LV networks

## 10.1 General

Comparison of low-voltage networks and distribution systems operating at higher voltage levels reveals a considerable number of similarities. Both are usually operated radially, and thus each network has only one infeed point. In the MV system the construction of a new 110/20 kV infeed substation can involve investment of the order of several million pounds and require extensive design work. However, such schemes are not so numerous as the provision of new 20/0·4 kV distribution substations, each costing only about £3000–£30 000, and the associated changes in operating arrangements of the low-voltage system. These latter schemes are carried out almost daily, depending on the size of the utility and the distribution network conditions. Although individual LV construction schemes are small, the large number of such jobs carried out each year tends to absorb a major part of a utility's capital and design resources. This leads to the need for good design practice involving an efficient organisation to oversee the large number of LV schemes, with standardised approaches to LV network design.

When planning for the electricity supply to a new area or the reinforcement of an existing network, it is necessary to consider the requirements for MV/LV distribution substations and LV circuits at the same time. For each scheme there are optimum locations and sizes for each substation, as well as the configuration and dimensioning of the associated LV systems. This topic is covered in more detail in Chapter 14, where computer-based design methods are discussed.

Figure 10.1 indicates how the costs of providing electricity supply are influenced by the number of customers fed from an individual distribution substation, the density of housing in the area being supplied, and the main construction policy which has been adopted, i.e. underground or overhead. The options most often used in system reinforcements are to replace existing

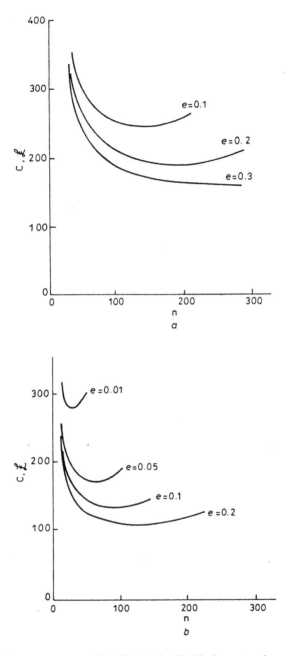

*Figure 10.1*  *Cost per consumer (C) of electricity distribution networks as a function of the number of consumers (n) fed by the transformer; service cables not included*

e: house area/plot area
*a* Underground cable network
*b* Overhead line network

overhead-line conductors by conductors of larger cross-sectional area and/or to add new MV/LV distribution substations. Here also it is necessary to combine studies to cover the effect of new substations and LV network changes.

In the United States general practice is to minimise the LV circuits, overhead or underground, by using more, and smaller, MV/LV transformers which may serve only a few customers. This arrangement is mainly due to the lower standard voltage used there.

If the ratio of the nominal voltages of two networks is 50:1, then, with the same conductor size, the higher-voltage network is capable of transferring 50 times as much power as the lower-voltage network for the same voltage drop. If the same percentage voltage drop is required, which approximates to present practice, then, at the higher voltage, a load 2500 times larger can be transferred over the same distance compared with the lower-voltage network. Alternatively, a load 50 times as great could be transferred over a distance 50 times as long. These conditions exist in 20/0·4 kV MV/LV systems. The normal power levels in LV circuits are therefore much lower than those associated with MV circuits. Consequently it is less economic to improve the security of low-voltage lines compared with that of medium-voltage lines of equivalent conductor size.

Low-voltage lines are more susceptible to lightning overvoltage than MV systems owing to their lower insulation level. Lightning strikes close to LV lines may induce overvoltages causing damage. The practice of keeping LV lines short, the present trend of placing systems underground, and the increasing use of aerial bunched conductors have proved successful in reducing these overvoltage problems.

The $R/X$ ratio for LV lines and cables is usually high. Thus reactive loads do not often cause voltage-drop problems (see eqn. 3.24), but the high $R/X$ ratio considerably effects the power losses (see eqn. 3.5).

Of necessity LV networks, services and appliances are often located in proximity to human beings and animals. The average person is not aware of the potential danger from the misuse of electrical appliances, or from the consequences of faults on their equipment or the installations at work, home or places of entertainment. The voltage levels used in Europe can lead to electric shock under adverse situations, so that strict Government and utility regulations are necessary to cover the safety aspects of wiring installations, protective devices within a customer's premises, and the individual electrical appliances. The degree of control imposed by such regulations has an influence on acceptable design arrangement and their costs.

There are large variations in load density between different areas. In the centre of large cities the average load density can exceed 100 MW/km$^2$ while in rural areas the maximum density may only be some tens of kilowatts per square kilometre, and less in remote areas. This leads to considerable differences in the optimum network arrangements. In rural areas small transformers, sometimes below 10 kVA rating and supported on poles, feed a few customers via overhead lines with the average load per line being sometimes only a few kilowatts. In city centres, substations at basement or street level inside commercial blocks, with a number of transformers in the 1 MVA range, feed many customers via strong underground cable networks.

## 10.2 Cable-connected substations

In urban areas where the load density is high the optimum distribution-transformer size tends to be between 0·5 and 2·0 MVA. Depending on the location, the MV/LV substation can be of a kiosk type or contained within a commercial building or a multi-storey housing block, often having to meet strict environmental and amenity requirements. In such areas it is usual for the MV and LV systems to be underground and, owing to the relatively high powers delivered through the MV/LV substations, reliability is particularly important. To ensure maximum continuity of supplies, looped MV networks as illustrated diagrammatically in Figure 10.2 are used.

Various switching and isolating arrangements are now in operation on underground urban MV systems. A diagram of a totally enclosed 'ring-main unit' is shown in Figure 10.3. The incoming MV feeders are provided with load-breaking disconnectors, with the transformer connected to the ring system through a loadbreak switch in series with a current-limiting fuse. The switch-fuse arrangement protects against a transformer fault and can disconnect a faulted

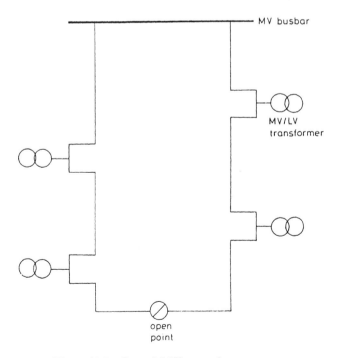

*Figure 10.2   Looped MV network arrangements*

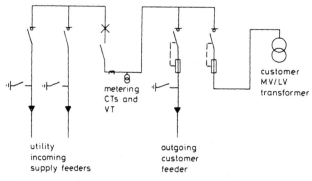

*Figure 10.6   Typical service arrangement for a large building*

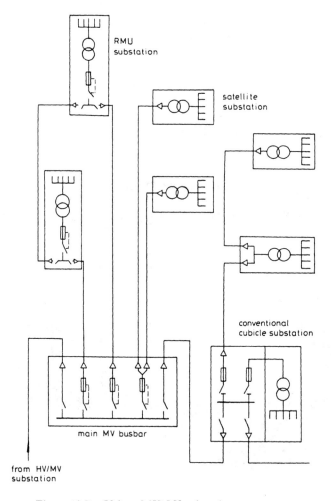

*Figure 10.7   Urban MV/LV substation arrangements*

considerable difficulties may be faced. The use of 'satellite' substations, each containing a single 100–300 kVA transformer, can be one solution. Cabled direct to the main MV infeed substation or into a convenient MV cable, as illustrated in Figure 10.7, the satellite substation requires no MV switchgear, being protected either at the main infeed substation or at some intermediate conventional MV substation; and it thus lends itself to a simple compact arrangement. Owing to their smaller size, suitable sites can be found more easily, and the use of satellite substations is becoming increasingly popular. A further advantage is that, being located near the load centres, this arrangement results in reduced LV circuit lengths.

## 10.3 Pole-mounted substations

In rural areas where the load density is low and overhead lines are used, pole-mounted substations are a natural choice. For MV/LV transformers below 100 kVA it is possible to bolt the transformer onto a single pole or place it on a metal frame attached to the pole. For larger-sized units a platform supported by two poles is used to hold the transformer up. Typical arrangements are shown in Figure 10.8, with some details omitted for clarity.

The use of individual MV fuses is not recommended since they frequently operate under storm or transient fault conditions. One 3-phase disconnector can cover a group of small pole-mounted substations connected to a section of the MV network, as discussed in Chapter 9. Larger transformers are equipped with individual disconnectors, as shown in Figure 10.8*b*. On the low-voltage side each feeder is usually equipped with its own protective fuses, although the use of miniature circuit breakers in place of fuses is increasing.

For overvoltage protection spark gaps are in common use on the MV side. A prerequisite for this arrangement is that the transformer-winding construction is not over-sensitive to the rapid collapse of surge voltages. Depending on local price levels and reliability appraisals and the average level of lightning strikes in the area, it may be economic to protect the larger rated transformers, say over 160 kVA, by lightning arresters. Further details of overvoltage protection are included in Section 7.6.

Pole-mounted substations also provide an economic arrangement in areas where MV overhead lines and LV underground cables are used.

## 10.4 LV network arrangements

Network practices vary as much as substation arrangements, depending very much on the district concerned. In urban areas with high housing density it is usually only possible to use underground cables for the LV system. In such situations LV cables from neighbouring distribution substations can terminate very close to one another, thus permitting interconnection of the LV network, at

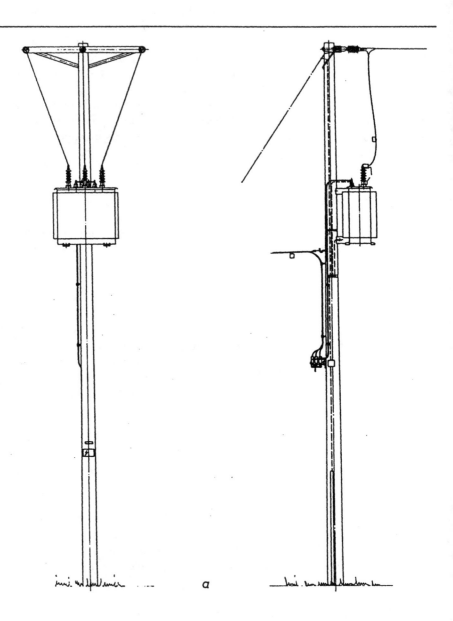

*Figure 10.8   Various methods of supporting transformers on poles (Courtesy Association of Finnish Electric Utilities)*

   *a* Single pole

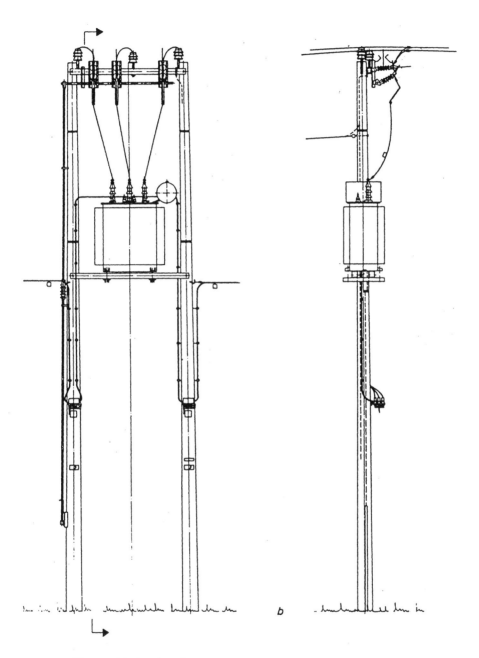

*Figure 10.8 (continued)*

*b* Twin poles

The basic target of dimensioning studies is to find the conductor size which minimises the total discounted costs of investments, losses, maintenance and operation over the life of the scheme, and also fulfils the necessary technical and safety constraints. The methods for determining the dimensions of MV lines given in Section 9.4 are basically valid also for LV line design. However, in this latter case the aspects which need to be emphasised, including safety regulations, are somewhat different.

In overhead-line circuits the most important constraints are voltage drop and adequate fault current to operate the protection. With short cables, thermal limits may lead to the rejection of otherwise acceptable alternatives, or at least reduce the availability of some cables for back-up supplies. The most appropriate ranges for two sizes of LV overhead cables are shown in Figure 10.10.

The vertical lines showing the limits of acceptable line length are derived from the need for adequate fault current compared with the feeder fuse rating. The curved lines are calculated from the maximum voltage drop. The continuous horizontal lines are based on the thermal limits, with the broken lines showing the economic effect of applying an interest rate of 5 or 10%. This type of diagram is only valid for a particular combination of assumed costs, load growth and specified safety regulations.

In practice, the load is not usually concentrated at the end of the feeder as was assumed in deriving Figure 10.10. The optimum cross-section therefore varies along the line, depending on the load level. This then results in a constant voltage drop per unit length of line. However, there is often only a small cost difference, for the same voltage drop, between networks with optimum conductor cross-section tapering and those using just one correctly chosen

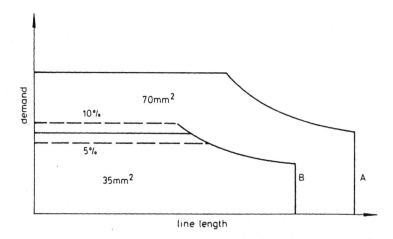

*Figure 10.10   Range of applications for various LV overhead cables, and their constraints*

   A 125 A feeder fuse rating
   B  80 A feeder fuse rating

standard-size cross-section conductor. Thus a maximum of two different conductor sizes in each LV network, correctly determined, will provide an economic arrangement in most cases.

## 10.5 Service connections

The method by which low-voltage customers are connected to the LV network depends mainly on the type of network (overhead or underground), the load density and local utility regulations. For example, the requirements for fault and overcurrent protection and the use of live-line working vary considerably in different countries.

In remote rural areas the low-voltage-supply conductors may be terminated on the walls of houses directly from the low-voltage overhead line. In other rural areas and small villages the service connection to a customer is usually provided by a spur which may have a lower cross-sectional area than the main line. Overload protection of the customer's installation can be provided by the main fuses at the customer's intake point. If local safety regulations also require protection for faults on the service, then fuses on the LV feeder at the distribution substation may not be sensitive enough, so that additional fuses may be required at the junctions of the overhead feeder and the services.

A service connection arrangement linking several customers is also accepted practice in some countries. In those countries where bare overhead conductors are still used, the service connection is often insulated. In more modern housing areas, where the buildings are close together and often of single-storey construction, the services are often underground although the LV mains are overhead. It is useful if the service joints to the feeders are so arranged that they can be connected and disconnected while the network is 'live', i.e. with voltage still applied. This enables the connection of a new house to be carried out without any outage of supply to the other customers connected to that feeder.

In housing estates there are many methods of connecting detached or terraced houses to the underground cable network, and some of these are shown in Figure 10.11.

In arrangement (i) the disconnection boxes not only provide facilities to connect up the services but also make it possible to provide fuses on each outgoing cable. Arrangements (ii) and (iii) are cheaper owing to the linking connections, but do not permit such individual good protection facilities as arrangement (i). Arrangement (iv) is for individual large or remote customers. Arrangement (v) represents the situation with fixed underground joints, which are much cheaper than cable disconnection boxes or cabinets although the selective protection facility discussed earlier is not possible.

Arrangements (vi) and (vii) are variations on arrangement (v). In particular, arrangement (vii) has proved to be cost effective in countries where each customer in a terraced house has his or her own service and this simple arrangement meets the local safety regulations.

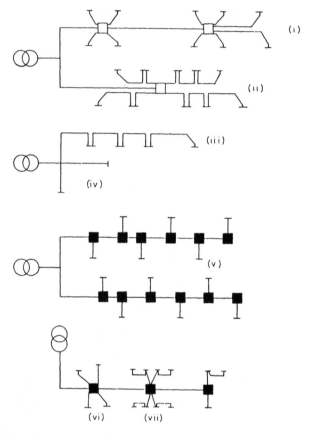

*Figure 10.11    Some service cable arrangements*

□ disconnection box
— customer
■ underground fixed joint

Determining the correct practices for service connections within a utility should be an optimisation process. Depending on the distribution network conditions, the practices adopted by one utility are not necessarily the most economic, or environmentally acceptable, under different circumstances. However, when general practices have been determined it is then possible to set out guidelines which can be easily applied by the design staff.

As an example of developing such guidelines consider the system shown in Figure 10.12, which is the same as arrangement (i) of Figure 10.11. The more customers connected to a disconnection box the longer and more expensive the services will become, but the cost of boxes per customer will decrease. The number of relevant factors involved makes it difficult to decide on the type and average distance between disconnection boxes for a particular new housing area,

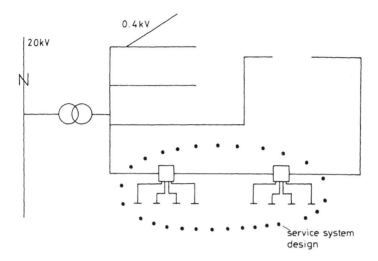

*Figure 10.12    Part of the distribution network considered in the service-system optimisation*

□ disconnection box
— customer

even if the general practice for feeder arrangements has been fixed. The voltage drop and costs of materials, excavation, joints and losses must all be considered when determining the size of the service cables.

An example of design guides obtained by computer program is given in Figure 10.13. The program utilises data files including plot-area and location information, the necessary calculation parameters, service-cable information

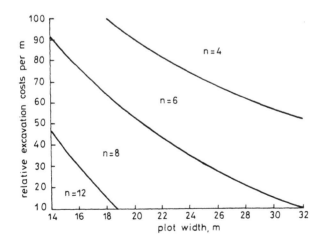

*Figure 10.13    Example of cost diagrams for service design*

*n* = optimum number of customers per disconnection box

customer cannot be determined precisely, it is usually necessary to calculate system loadings on a statistical basis, whether considering existing loads or forecast values.

## 11.2 Load-monitoring measurements

Load data are collected from an existing network for circuit monitoring, for tariff purposes, and for various research and supply quality-control studies. For simplicity the tariffs of most customers do not require demand or load curve measurements, so that these must be obtained separately for the above-mentioned studies. It is therefore essential to build in a system to transmit all the relevant measurement data to the utility's load file.

Telecontrolled substations offer an ideal arrangement for collecting transformer and feeder load data. However, if the interval between load sampling is too short, this can lead to large amounts of data being transmitted which cannot always be processed or utilised sensibly, unless steps are taken to select only essential data. Large distribution substations are often equipped with a maximum-current meter, maximum-demand (kVA) indicator, or a kWh meter. With advances in distribution-system automation these measuring devices are being connected into supply-network data-collection schemes, which sometimes include data transmission to and from the substation. Occasionally load and voltage measurement surveys are carried out in specific LV networks in order to check the quality of supply.

The short-term aim of the above measurements is to ensure that the existing distribution system is operating satisfactorily, in order to determine where rearrangement of the feeder configuration could improve the overall system performance, and also to locate any areas where improved system performance is likely to be required in the next few years. In addition to the above, it is essential to collect load-curve information since this can be used when developing the load forecasts necessary for both the long-term, and the more detailed short-term, system-design work.

As discussed in Section 11.3, the instantaneous customer and system loads are often considered to have Gaussian distributions. In order to develop the relevant mean values and standard deviations of load curves for customers belonging to difference classes, large numbers of recorded consumption values are required. There must be at least 100 metered customers for each customer class with consumption records taken over the last three years; or a smaller number for a longer time period.

When setting up a measurement project to obtain the above data, alternative means of obtaining the load data must be carefully assessed. In addition, it is useful to obtain background information on each series of tests and for each metering point, e.g. customer class and data on ambient temperatures throughout the test, so that these factors can be taken into account when finally analysing the studies. Within the relatively small area covered by an LV network, knowledge of the maximum demand at each substation and the unit consumption can provide sufficiently accurate data for statistical analysis to produce worthwhile load forecasts, even if data on individual LV circuits are not available.

## 11.3 Load analysis and synthesis

Where full loading information is not available from metering equipment on the system, techniques of load analysis and synthesis may be used, but they should, where possible, be checked against measured information. The power demand of any one customer has a daily pattern which is influenced considerably by the day of the week concerned and also the time of year. Daily and seasonal load curves can be constructed for various customer groups if satisfactory metered data are available. The distribution of the load for a particular hour of the day can be approximated by a Gaussian distribution curve, which is characterised by its mean value and standard deviation.

Load curves can then be used to simulate the load of one customer belonging to a particular customer group by scaling the annual units of that customer to the average units of the group. A load curve for a group of customers can be constructed by summing the individual customer load curves statistically. If required, load curves can also be constructed using abnormal values for some of the parameters involved, e.g. by drastically changing the input value of the ambient temperature to simulate extreme winter or summer conditions.

There are considerable variations in the load curves and the standard deviations for different types of customers. Figure 11.1 gives examples of the variations in mean demand over 26 two-week periods for a customer with electric space heating, a domestic customer without electric space heating, and an agricultural customer.

Distribution utilities often categorise customers into different classifications for billing purposes, statistical analysis etc. In assessing whether such a classification is valid for generating load curves, the following points should be borne in mind:

- The main features of the consumption pattern of individual customers within a given customer class must be very similar, otherwise the standard deviations will become excessive.
- The number of classes should not be too high, usually below 15, in order to keep the costs of data collection and manipulation within reasonable limits.
- The customer data system should be related to the classification codes of the utility's customer information system. This will lead to increased use of automatic methods of utilising load information for distribution system studies.

Some examples of customer classification are given in Table 11.1. Other examples might include commercial- and industrial-customer consumption information, with perhaps the industrial figure split into groups having 1-, 2- or 3-shift working each day, and street lighting.

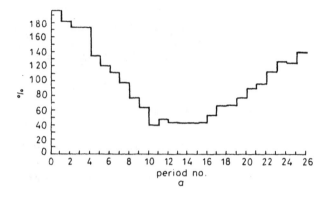

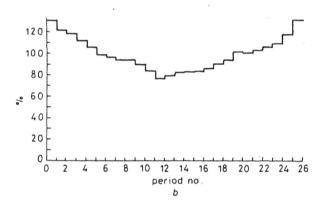

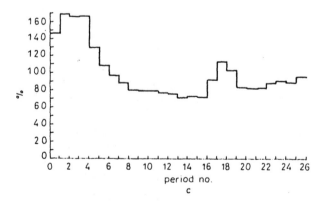

*Figure 11.1    Examples of mean demands for different types of customer*

100% ≃ annual mean
*a*  Electric space heating
*b*  Domestic
*c*  Agriculture

Usually the actual load curves at a customer's service infeed point, or at various points on a network, are not known. The energy supplied to each customer can, however, be obtained from the billing files for existing conditions, and estimates made for future conditions. If practically all the customers in the area under study belong to one particular customer group, then approximate peak demands can be estimated by more conventional methods such as the use of load factors, or the use of Velander's formula for this particular group as given by

$$P = k_1 W + k_2 \sqrt{W} \tag{11.1}$$

where $P$ is the peak demand of the group, $W$ is the annual unit consumption and $k_1$ and $k_2$ are constants.

*Table 11.1 Examples of Velander coefficients*

| Customer group | $k_1$ | $k_2$ |
|---|---|---|
| Domestic | 0·29 | 2·5 |
| Electric space heating | 0·22 | 0·9 |
| Commercial (shops) | 0·25 | 1·9 |

Typical values for the constants in eqn. 11.1 for some customer groups are given in Table 11.1. With $W$ given in MWh, $P$ is obtained in kW.

For areas containing a number of customer groups, a statistical method should preferably be used to sum the load curves for the different customer groups.

For each time period the demand of each customer is represented by a mean value and a standard deviation. The summed mean value of the demand over any one hour is the sum of the mean values of the individual component loads. The standard deviation of the summed load depends on the correlation between the component loads. If there is no correlation between them, the total standard deviation is given by

$$\delta_t = \sqrt{\sum \delta_i^2} \tag{11.2}$$

The percentage deviation is obtained by dividing $\delta_t$ by the sum of the mean values of the loads. Thus, in this case, the percentage deviation decreases as the number of customers increases. If there is a positive correlation between customer demands, the total deviation decreases more slowly.

In Figure 11.2 the lower curve 1 represents the mean values for a terraced house with four domestic customers (total sum is 20 MWh/year) in kilowatts over a 24 h period. The upper curve 2 will be exceeded with a small 1% excess probability.

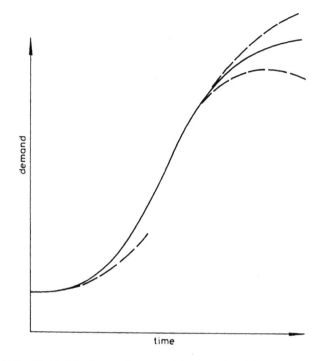

*Figure 11.5   Typical load-growth curve with extrapolated forecasts at varying times*

trend usually follows an S curve as shown in Figure 11.5. In the earlier years of an electricity-supply system, there may be a high annual rate of growth in demand, but as individual customers' dwellings become saturated with electrical equipment the rate of increase will drop as indicated by the full line in Figure 11.5. Economic factors within the area concerned and in the country overall, as well as in the influence of relative costs of alternative energy sources, also have an influence on the growth rates.

Considerable errors can occur in the forecast since this method takes no account of the stage of development of the area under review.

*Simulation methods* are based on using specific annual consumption figures, obtained from surveys of individual consumption classes and the number of customers in each consumption class. The present situation can be printed out from the utility billing files. The future development of each consumption class can be estimated from national information and then modified for local use, e.g. by information from municipal or parish authorities on expected future industrial and housing developments. Simulation forecasts are especially relevant for areas where large developments are expected or when the forecast period is longer than five years or so.

*Econometric modelling* is based on obtaining suitable correlation between power consumption and various economic parameters such as gross national product, index of industrial production and rate of inflation. The effect of the price of electricity is also relevant. Large-scale econometric models are not appropriate for the small areas usually associated with individual distribution-system studies.

Load forecasts at individual HV/MV substation level can often be based on both of the first two methods mentioned above. The existing and predicted future population being supplied, and likely future housing and industrial developments, can be assessed from discussions with the appropriate local area-planning authority, using the billing file to provide information on existing and past consumptions as a starting point for such load forecasts.

If the expected development of each customer group, and any increase in the assumed consumption within that group, are taken account of a forecast can then be made of the predicted future consumption of the group.

Table 11.2 is based on a survey carried out in Finland. The first line, number of houses, is the sum of the family houses, with and without electric space heating, and farms. Considering the first customer group covering houses with

*Table 11.2  An example of forecasting area consumption*

| Year | 0 | 10 | 20 |
|---|---|---|---|
| *Population* | 11 700 | 11 900 | 12 200 |
| *Number of houses* | 4550 | 4940 | 5450 |
| % with electric space heating | 14 | 21 | 32 |
| Number with electrically space-heated houses | 637 | 1037 | 1744 |
| Heating consumption, MWh/house per year | 17·1 | 17·6 | 18·1 |
| Consumption, MWh/year | 10 900 | 18 300 | 31 600 |
| *Number of family houses* | 4135 | 4540 | 5080 |
| Domestic consumption, MWh/house per year | 4·2 | 4·7 | 5·2 |
| Consumption, MWh/year | 17 400 | 21 300 | 26 400 |
| *Number of summer cottages* | 1030 | 1240 | 1600 |
| Number electrified | 490 | 650 | 930 |
| Specific consumption, MWh/cottage per year | 1·5 | 2·5 | 3·0 |
| Consumption, MWh/year | 740 | 1630 | 2790 |
| *Number of electrically space-heated summer cottages* | 200 | 330 | 560 |
| Heating consumption, MWh/cottage per year | 4·2 | 4·6 | 5·0 |
| Consumption, MWh/year | 840 | 1500 | 2800 |
| *Number of farms* | 415 | 400 | 370 |
| Specific consumption, MWh/farm per year | 5·6 | 6·7 | 8·0 |
| Consumption, MWh/year | 2300 | 2700 | 3000 |
| *Number of industrial workers* | 1350 | 1400 | 1475 |
| Specific consumption, MWh/worker per year | 6·1 | 7·0 | 8·2 |
| Consumption, MWh/year | 8200 | 9800 | 12100 |
| *Number of public and services workers* | 2650 | 2740 | 2800 |
| Specific consumption, MWh/worker per year | 5·4 | 5·9 | 6·8 |
| Consumption, MWh/year | 14 300 | 16 200 | 19 000 |
| *Total consumption,* MWh/year | 54 700 | 71 500 | 97 700 |

electric space heating, then, by using the forecast number of houses, the percentage of these houses expected to have electric space heating, and the predicted increase in consumption per house each year, the total consumption for this customer group can be forecast as rising from 10900 MWh in year 0, which is the last year for which statistics are available, to 31600 MWh by year 20. Similar calculations are used to derive the forecast consumption by cottages and farms. Energy sales to industrial customers have been based on forecast electrical consumption per worker in specific industrial groups linked to the predicted number of workers in each group in the years under consideration. The total forecast consumption for the area can then be obtained by summing the individual group forecasts as shown.

The total area consumption is obtained by summing the individual customer-class consumptions, as shown in Table 11.2. Such calculations can be carried out by suitable computer programs and the results used directly in network-design programs. The energy forecasts obtained can be converted to demand forecasts by applying the Velander formula of eqn. 11.1 for each customer group, and taking into account the differences in time for the peak demands for each group.

Load forecasts at distribution substation level (MV/LV) can be based on the above-mentioned MV studies, in many cases extracting information on computer file relating to customer loads, network system data, and known developments if these are part of the computer filing system. It is also possible that customer-appliance information for specific substation feeding areas may reveal parts of the area where more rapid load growth may occur, compared with the average situation. An example of this is electric space heating in rural areas. Since this can be a popular heating alternative, those areas with a lower percentage of customers with electric space heating have potential for rapid load growth, which should be kept in mind when developing the associated electrical systems.

Load forecasts for low-voltage networks in urban areas can be generated in the same way as those for distribution substations described in the previous paragraph. In rural areas the use of average values of load growth is not recommended. Here the main part of load growth comes from individual new customers, or by a transfer to electric space heating, which can have a large effect on one or two local circuits but not affect the distribution network outside the area local to such a load increase. Therefore, to apply the principle of estimated average load growth could lead to too high loading estimates for most lines, and much too low estimates where new customers are likely to be connected, or where existing heating is changed to electric. Unlike individual LV circuits the MV feeders supply hundreds, and often thousands, of customers. Thus the average load-growth forecasts for the area being supplied are appropriate when carrying out MV network design.

It is essential that load forecasts can be relatively easily adjusted when any new background information becomes available. It is therefore worthwhile that there should be close liaison between those responsible for the data banks covering load forecasting, customer information and network data. The cross-indexing of all the interrelated data is vital for optimising the load-forecasting process, which is an essential stage in the network planning process. This means, for example, that similar location codes and customer classifications are used in all these systems and that good communications exist with community planning staff.

## 11.5 Short-term load forecasts

While the type of long-term forecast introduced in the previous Section is relevant for network planning purposes, short-term forecasts are needed for the operation of distribution systems, with typical lead times ranging from one hour to one week.

New automation facilities such as remote controlled disconnectors and microprocessor-based relays, together with low-priced computer technology, have made it possible to develop systems that support distribution network operation in a more sophisticated manner, as discussed in Section 7.7. These distribution management systems (DMS) include real-time network analysis and support the operator in operation planning. DMS usually necessitates the integration of a utility's computer systems. The supervisory control and data acquisition system (SCADA) provides real-time data from primary substations and various devices in the network. On their own, the data obtained from the SCADA are usually not adequate for real-time monitoring of the state of the distribution network as a whole, since detailed information from lines and network components, and the load distribution along MV feeders and loads on various spurs, are not known. However, the data on network components which are required are available in a network information system (AM/FM-system), which will be introduced in Chapter 14.

The advanced functions of distribution network operation, for example restoration and power-loss reduction by feeder reconfiguration, require load forecasts in order to achieve accurate results. For these functions the forecast results must be flexible enough to be used in any simulated network configuration, so that the target of the short-term load forecasting method for distribution network operation is to forecast the loads of the distribution substations (20/0·4 kV). The load forecasting method is based on combining available remote load measurements with the load models of customer groups introduced in Section 11.3. The present load on each distribution substation is calculated based on load models of various customer groups, and on the prevailing weather information. These load models include hourly demands and temperature dependencies for, say, 50 customer groups. The modelled loads of distribution substations together with the voltage measured at the feeding-point busbar are used in the load-flow calculation of the medium-voltage network to arrive at the first estimate of the state of the network.

The principles of state estimation are then applied, with redundant and doubtful measurements being used in a statistical process utilising, for example,

*Chapter 12*
# Special loads

## 12.1 General

Customers expect an electricity supply of good quality. Consequently it is necessary to give special consideration to loads which may produce various irregularities on the supply voltage, resulting in interference with the correct operation of customer appliances or utility equipment. Typical of such loads are steel-making arc furnaces, welding equipment, induction furnaces, rolling mills and colliery winders, and railway traction, where rapid variations in load currents may result in fluctuations in the voltage at customers' intake points. While the larger industrial loads will often require individual attention, there are also items of equipment, mainly in use in domestic and commercial premises, which, while individually not causing problems, can collectively affect the quality of supply owing to the large number of items involved. In addition some installations, such as computers and process-control equipment, are themselves susceptible to the quality of the supply voltages.

The overall effect of these 'disturbance loads' on individual supply voltages will depend on such factors as the magnitude, phase angle and rate of change of the currents taken by the load, and whether the load changes occur at regular or random intervals of time. The frequency of such load changes and whether they occur at time of peak demand, or at off-peak periods such as during the night, have a bearing on their interference with the operation of other equipment.

## 12.2 Electric-arc furnaces

### 12.2.1 Load characteristics

A particular feature of the operation of electric arc-furnaces is the frequent recurrence of short circuits between the electrodes and the scrap-metal charge.

Often when the molten scrap metal drops away from an electrode the arc will extinguish and no current will flow. During the melt-down period there will thus be random current changes with two or three phases short-circuited, or one phase on open circuit. The swings from short circuit to open circuit produce violent current fluctuations, often several times larger than the furnace nameplate rating, whether this be some kilowatts or tens of megawatts. These result in large voltage variations being impressed on the incoming supply voltage, usually LV for the low ratings and MV for the MW range of furnace ratings. Since the fluctuating load current to the furnace passes through the supply network, a corresponding fluctuation is impressed at the busbar linking other customers on the same network, which is often referred to as the *point of common coupling* (PCC), as indicated in Figure 12.1.

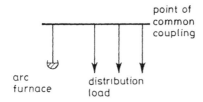

point of
common
coupling

arc
furnace

distribution
load

*Figure 12.1   Point of common coupling*

## 12.2.2  Voltage fluctuations and lamp flicker

The fluctuation of the supply voltage which occurs during the melt down of the scrap metal can cause flicker in incandescent lamps, to which the human eye is very sensitive. Tests have been carried out in a number of countries to assess what amount of voltage variation can be tolerated by customers in order that the electricity supplier can limit the voltage fluctuations to a level where unacceptable flicker is avoided. The sensitivity of the human eye to lamp flicker, relative to changes in the supply voltage, is illustrated in Figure 12.2.

As a result of the movement of the arc between the electrodes, the furnace current rapidly fluctuates at varying frequencies from below 0.1 Hz to in excess of 10 Hz. This causes an irregular fluctuation of the supply voltage. A section of this fluctuation is shown in Figure 12.3$a$, where the supply voltage waveform, having an instantaneous value $v$, may be considered as a carrier wave being modulated by the random voltage fluctuations caused by the arc furnace. This modulation voltage may then be considered as being independent of the normal supply waveform but having an instantaneous value $v_f$ around its own notional zero line, as shown in Figure 12.3$b$.

It is necessary to derive a single parameter of this fluctuating voltage, so that this can be measured and used to define the severity of flicker to human subjects. Tests have shown that the RMS value of the voltage fluctuation $V_f$ is a

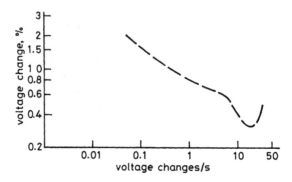

*Figure 12.2   Maximum value of voltage change to avoid flicker annoyance*

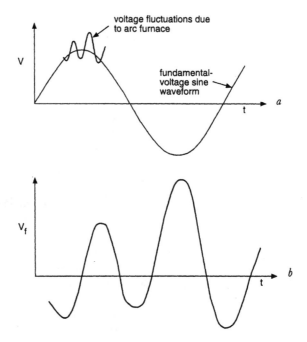

*Figure 12.3   Voltage fluctuation*

        *a* Voltage fluctuation impressed on fundamental voltage waveform
        *b* Fluctuation voltage around fundamental voltage

satisfactory measurement for arc-furnace voltage flicker, and this value can therefore be used as a measure of flicker severity to the human eye. In practice, $V_f$ is quoted as a percentage of the RMS value of the supply voltage, typical values being some tenths of 1%.

The short-circuit voltage depression $V_t$ is the percentage change of voltage at the point of common coupling when the furnace electrodes are taken from open circuit to short circuit by dipping them into the molten metal. $V_t$ can be calculated to within reasonable accuracy from

$$V_t = \frac{S_f}{S_{pcc}} \times 100\% \qquad (12.1)$$

where $S_f$ is the apparent short-circuit power of the arc-furnace as seen at the point of common coupling and $S_{pcc}$ is the fault level at the point of common coupling. In order to arrive at a suitable value of $S_{pcc}$ it is appropriate to carry out calculations based on system operating conditions when the infeeds to a short circuit are low, e.g. due to reduced generation at time of light load, or because of the loss of an EHV/HV transformer feeding into the system.

For a group of arc furnaces, either at the same installation or in the same locality, or of sufficient power to transmit flicker via the EHV networks to the point of common coupling under consideration, it is necessary to derive the equivalent short-circuit furnace MVA to the point of common coupling. For a number of furnaces having apparent short-circuit powers $S_{f1}$, $S_{f2}$, $S_{f3}$ etc, this equivalent short-circuit power $S_{eq}$ is calculated as the $n$th root of the sum of the $n$th power of the short-circuit powers of each arc-furnace at the point of common coupling; i.e.

$$S_{eq} = \sqrt[n]{(S_{f1}^n + S_{f2}^n + S_{f3}^n + \ldots S_{fi}^n)} \qquad (12.2)$$

and the combined flicker $V_{FG}$ at a given point of common coupling can be assessed from

$$V_{FG} \simeq \sqrt[m]{\sum V_{fg}^m} \qquad (12.3)$$

where the flicker contribution $V_{fg}$ from any installation under review, plus those from existing installations, is defined as a percentage value of the supply voltage at the point of common coupling. $m$ can be between 2 and 4 depending on the operating mode of each furnace affecting the installation being assessed.

Alternative methods of defining the severity of voltage fluctuations are given in Section 12.7.2.

## 12.2.3 Methods for reducing voltage fluctuation

Where excessive voltage fluctuations are caused by arc-furnace installations so that supplies to other customers are seriously affected, or calculations indicate that a proposed arc-furnace installation will result in excessive voltage fluctuations, a number of options are available to reduce arc interference. These may involve re-arranging the system configuration to minimise the effect of the arc furnace on other customers, or adding some compensation devices to counteract the arc-furnace reactive-power swings.

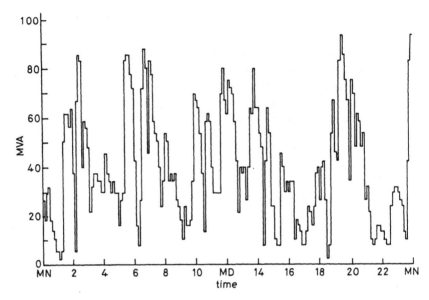

*Figure 12.6    Multiple-arc furnace load cycle (Courtesy Midlands Electricity plc)*

Consider the four sets of load currents varying in time around a 1.0 p.u. level, as shown in Figure 12.7a: (i) the $I^2t$ effect for a load current of 1.0 p.u. for 1.0 p.u. time is $(1 \cdot 0)^2 \times 1 = 1 \cdot 0$; in (ii) the current variation is $\pm 0 \cdot 2$ p.u., also over $1 \cdot 0$ p.u. time, and the $I^2t$ effect for this case is then $\frac{1}{2}(1 \cdot 2)^2 + \frac{1}{2}(0 \cdot 8)^2 = 1 \cdot 04$ p.u. Thus although the average current in this case is still 1.0 p.u., the heating effect has increased by 4%.

Similarly for two equal current swings up to $1 \cdot 4$ p.u. and two down to $0 \cdot 6$ p.u., shown in Figure 12.7a (iii), averaging 1.0 p.u. over a $1 \cdot 0$ p.u. time, the net heating effect is $2(1 \cdot 4)^2/4 + 2(0 \cdot 6)^2/4 = 1 \cdot 16$. For four swings up to $1 \cdot 8$ p.u. and four swings down to $0 \cdot 2$ p.u., as in Figure 12.7a (iv), the average current is still 1.0 p.u. but the heating effect is $4(1 \cdot 8)^2/8 + 4(0 \cdot 2)^2/8 = 1 \cdot 64$, and this trend is shown in Figure 12.7b.

Thus for the widely varying short-time current swings experienced with arc-furnace loads, the actual cumulative $I^2t$ effect is considerably greater than would be expected from the average half-hourly demand values. The cyclic rating of the supply transformer, as discussed in Section 6.2, is therefore inappropriate. Depending on the actual load cycle it may be necessary to install a transformer with a nominal rating at least equal to the predicted maximum half-hourly average demand or up to 20% higher. In addition, the transformer manufacturer should be informed of the nature of the load being supplied through the transformer, as special bracing of the core and windings may be necessary owing to the electromagnetic stresses induced by the heavy currents during the melt-down period of arc-furnace operation.

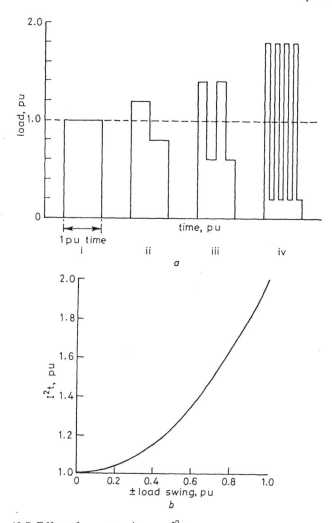

*Figure 12.7 Effect of current swings on $I^2 t$*

    *a* Varying load patterns for $I^2 t$ calculations
    *b* Variation in $I^2 t$ effect with load swings

## 12.3 Convertors

In this Section the harmonic currents and voltages caused by rectifiers and inverters are discussed. Rectifiers convert from AC to DC and inverters from DC to AC. The use of heavy-current rectifiers and inverters has contributed significantly to the development of variable-speed DC and AC motor drives, which have many applications in industry and rail traction. However, these installations can be sources of harmonic distortion affecting other customers.

An ideal diode allows current to flow in one direction only, the current being determined by the supply voltage and the load impedance. With reversal of the input voltage the diode acts as an infinite impedance and no reverse current flows. Considering the arrangement shown in Figure 12.8a, with a diode connected into each of the three phases of an AC supply, reference to the voltage waveforms in Figure 12.8b shows that at any point in time one or two diodes will have a positive AC voltage applied to them. Conduction will take place through the diode subjected to the higher voltage.

As the voltage in one phase falls and that in another rises a point will be reached when the voltage across two units will be equal, as at $ab_1$, $bc_1$ and $ca_1$. Conduction will then pass from the unit receiving the falling voltage to the unit

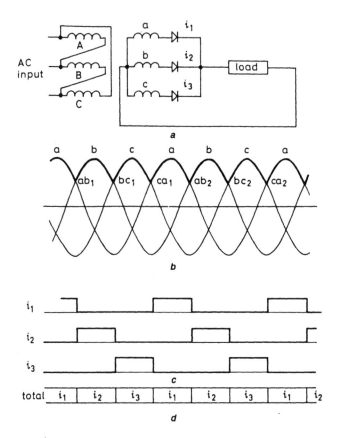

*Figure 12.8*   *3-phase rectification*

       *a* Schematic of connections
       *b* Input voltage/current waveforms
       *c* Ideal input-current waveforms
       *d* Ideal output-current waveforms

receiving a rising higher voltage. This transfer of conduction is called commutation. Individual outputs occur during time $ab_1 - bc_1$, $ab_2 - bc_2$ etc. in one phase, and similarly in the other two phases but displaced by 120° and 240°. As shown in Figure 12.8d this results in a continuous DC output under this theoretically ideal situation, where the load has been assumed to be purely inductive and, seen from the AC input of the convertor, each phase is loaded by currents which have a rectangular shape. Since these pulses are no longer pure sine waveforms, harmonics appear. For a particular harmonic current flowing through the supply network, a voltage drop will be produced in the system series impedances, and thus a harmonic voltage is produced. The summation of these individual harmonic voltage drops gives the total harmonic voltage distortion.

The harmonic currents also cause increased losses, thus decreasing the loading throughput capacity of a network. They may also cause errors in energy meters and system protection. Serious problems can occur if the frequency of one harmonic coincides with the resonant frequency of the network, resulting in overvoltages. In addition, harmonics can result in increased vibration in transformers and motors.

The arrangement shown in Figure 12.8a is the simplest of 3-phase rectifiers, and is known as a 3-pulse rectifier because there are three individual pulses of current ($i_1$, $i_2$ and $i_3$) in the DC output for each complete cycle of AC input. An improvement on this arrangement is to provide two diodes in each phase, as shown in Figure 12.9a, so that full-wave rectification takes place. An analysis of the individual currents will show that six pulses of direct current are produced during each complete cycle of AC input, and this arrangement is known as a 6-pulse rectifier.

From the point of view of suppression of harmonics, the 6-pulse bridge is the best that can be obtained with one 3-phase AC input. To obtain further improvement more phases are required, and a special transformer must be used. For example, a transformer with two secondary windings, one connected in star and the other in delta, will provide six phases spaced 60° apart. With each secondary winding connected to a pair of rectifiers, as shown in Figure 12.9b, the DC output now has 12 pulses for each cycle of AC input. Similarly 24 pulses, and even higher numbers, can be achieved. The reasons for going to these higher pulse numbers will be appreciated when consideration is given to the harmonics generated by the various arrangements.

The harmonics generated by an ideal rectifier are given by

$$N = kp \pm 1 \tag{12.4}$$

where  $p$ = pulse number

$k$ = any integer from 1 to infinity

$N$ = harmonic number

Thus the simple 3-pulse rectifier produces all harmonics, except the triplens where the harmonic is a multiple of 3. With a 6-pulse arrangement, the even

## 12.4 Motors

There are three main problems which can arise from the connection of motors to electrical supply systems. The first is whether, under depressed voltage conditions at starting, the motor will successfully run up to speed. For larger motors the customers will need details of the system characteristics to check this. The second results from the effect on other customers of this voltage depression when starting the motor from standstill. The third arises from the currents which motors feed back into the network when the supply voltage is suddenly reduced because of faults on the network.

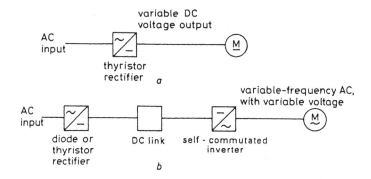

*Figure 12.11 Speed control of motors*

### 12.4.1 Starting currents

Switching any load which has a demand greater than about 0·25% of the network short-circuit fault current can cause disturbance to other connected loads. The degree of interference will depend not only on the magnitude of the current taken but also on its phase angle relative to the system voltage, and on whether the current change is gradual or sudden, and on how frequently the change occurs. In addition, variations which occur frequently tend to be more disturbing to other customers.

Direct on-line starting of a motor at standstill is a particular case of this, when the current will be several times the full-load motor current and at a relatively low power factor, typically 0·3 or less. The initial current is limited only by the system impedance and the internal impedance of the motor. As the motor runs up to speed the current will be reduced because of the back EMF generated by the motor, and its power factor will increase.

The initial current of a squirrel-cage induction motor, if started direct on line, will typically be 5–8 times full-load current, although special double-cage motor designs can reduce the current to 3·5–5 times full load. Star–delta starting conditions can be used to limit the initial current to within the range 2·5–3·5 times, while an auto-transformer start could further limit this to within 1·5–3·5 times full-load current. Thyristor-controlled soft-start devices are also popular. With these arrangements the starting current is automatically kept virtually constant. More expensive wound-rotor machines employing resistance starting can reduce the initial current to 1·5–2·5 times full-load current. Although some stepped starting methods, such as star–delta, may have higher current changes at changeover, this will be at a much higher power factor and the resulting voltage depression will usually be less than the voltage depression at switch-on.

## 12.4.2 Voltage variations

The percentage voltage variation caused by motor starting can be calculated from

$$V\% \simeq \{\sqrt{3}I/V\} \times (R\cos\phi + X\sin\phi) \times 100\% \qquad (12.6)$$

where  $I$ = motor starting current

$R$ and $X$ = resistive and reactive components of system impedance

$\cos\phi$ = power factor of motor starting current

$V$ = system phase–phase voltage

Care must be taken to use appropriate values for $R$ and $X$ in calculating the voltage variation for single-phase motors, allowing for feed and return paths.

UK experience indicates that typical limits to avoid unacceptable interference to other customers' equipment would be 1% of phase voltage at the point of common coupling for sudden voltage changes caused by frequent starting, and up to 3% where voltage changes are gradual over 2 s or more. The limit for sudden changes may be increased to, say, 3% if starting is less frequent than every 2 h, and up to 6% if starting is only once or twice a year. If several motors, which cause voltage depressions close to the limit of acceptability, are connected to the same point of common coupling, the combined effect should be considered, and where they involve frequent starting the limit may have to be reduced in accordance with Figure 12.2.

## 12.4.3 Short-circuit contribution

The contribution to short-circuit levels by motors can be considerable in the first half-cycle, and may still be appreciable at time of circuit-breaker contact separation with fast clearance times. The contribution from synchronous motors to the short-circuit current is more significant because the decay time of these machines is longer. The value can be calculated in the same way as for synchronous generators, as discussed in Chapter 3.

The contribution which induction motors can make has not always been appreciated, particularly in respect of the capability of switchgear when closing onto a fault. In general, a supply utility will not have specific information about the number, sizes or characteristics of the many small motors installed in customers' premises. These motors can therefore only be dealt with in groups, or in terms of the proportion of motor load to total load on the network. The initial contribution to short-circuit current from these small motors can be up to seven times their full load rating, and in total will depend on the proportion of connected motor load to total load. Typically figures are 1 MVA of fault infeed per MVA of aggregate non-industrial load, and up to 2.6 MVA per MVA of aggregate industrial load. The time constant of decay for small induction motors is about 30 ms.

When considering induction motors, the initial AC and DC components of fault current will have waveforms similar to those shown in Figure 12.12a, with the AC component remaining steady and the DC component decaying exponentially with time. The instantaneous value of current at time $t$ after fault initiation $i$ is given by

$$
\begin{aligned}
i = {} & \sqrt{2}\, I_{rms} \sin(\omega t + \theta_1 - \theta_2) \\
& - \sqrt{2}\, I_{rms} \sin(\theta_1 - \theta_2) \exp\{(-R/X)\omega t\}
\end{aligned}
\tag{12.7}
$$

where $I_{rms}$ = root-mean-square value of the AC component fault current

$\quad\;\; \theta_1$ = closing angle which defines the point on the source sinusoidal voltage when the fault occurs

$\quad\;\; \theta_2$ = system impedance angle, $= \tan^{-1} X/R$

$\quad\;\; \omega$ = angular frequency, rad/s

From eqn. 12.7, the peak value of the AC component is $\sqrt{2}$ times the RMS value of the AC component. If the closing angle is such that $\theta_1 - \theta_2 = n\pi/2$ radians, where $n$ is an odd integer, e.g. 1, 3, 5 etc., the DC component will have a maximum value at $t = 0$, immediately the fault occurs, of $\sqrt{2}\, I_{rms}$ also. The combined AC and DC component will then total twice $\sqrt{2}\, I_{rms}$, the so-called 'doubling effect'.

In Figure 12.12a the current waveforms have been drawn for an R/X ratio of 0·1, and a closing angle $\theta_1$ of 0°. The ratio of the maximum peak current to the RMS value of the AC component, the peak current factor, depends on the instant of the fault and the rate of decay of the DC component. The maximum peak current occurs when the closing angle $\theta_1$ is zero, irrespective of the $R/X$ ratio of the system. Figure 12.12b gives the peak current factors for a range of $R/X$ ratios.

For the larger induction motors, say those with a rated output exceeding 1% of the short-circuit fault level of the network under consideration, AC and DC components can be individually calculated using the locked-rotor reactance. In assessing the breaking duty of switchgear, allowance should be made for the

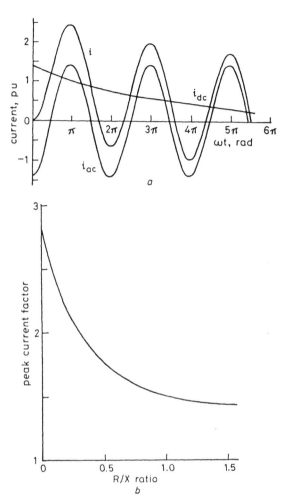

*Figure 12.12* Asymmetrical fault-current waveforms, and variation of peak current with R/X ratio

a Plot of total (asymmetrical) current for a 0° closing angle, $R/X = 0.1$. Total current consists of a DC component $i_{dc}$ and an AC component $i_{ac}$
b Variation of peak current with $R/X$ ratio

reduction in peak current with time, as both AC and DC components decay much more rapidly with induction motors than with synchronous machines. In practice, the contribution from most induction motors, unlike synchronous machines, may be disregarded after 100 ms. This is because those motors which have not tripped due to under-voltage will either have reverted to motoring at reduced supply voltage or will be providing infeeds out of phase with the infeeds from the rest of the system.

The infeed from the 132 kV busbar is the combined infeeds from the source and the industrial motors, represented by 0·067 Ω and 0·548 Ω impedance, respectively; i.e. equivalent to $(0·067 \times 0·548)/(0·067 + 0·548) = 0·060$ Ω impedance. Taking the impedance between the busbars at 0·608 Ω at 11 kV the infeed from the 132 kV busbars is then

$$2·75\frac{11}{\sqrt{3} \times (0·608 + 0·060)} = 26·2 \text{ kA}$$

giving a total infeed to the 11 kV busbars of $5·27 + 26·20 = 31·47$ kA.

As mentioned earlier, when considering the breaking duty on switchgear, the smaller motors will produce negligible contribution after 100 ms and the contribution from the two larger motors in this example would be considerably reduced.

## 12.5 Railway traction

It has been forecast that, by the end of the 20th century, some 40% of the world railway traffic routes will be electrified and carrying 75% of the total rail traffic. While some railway traction operates on DC using 3-phase trackside rectifiers, mainly in urban or suburban areas, the majority of traction supplies are now being provided at single-phase AC, typically 25 kV 50 Hz.

Particularly when starting heavy train loads, large currents are drawn from the infeed supply points. Figure 12.14 shows a typical recording of a main-line station where the starting of peak-hour inter-city trains, and of local smaller units, can be distinguished.

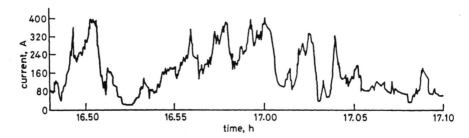

*Figure 12.14   Current recording at traction-supply point (Courtesy Midlands Electricity plc)*

There has been an almost total use of diode rectifiers or thyristor converters on traction power units in order to use DC motors. As discussed in Section 12.3, diodes and thyristor arrangements introduce harmonic distortion into the supply network. Provision of single-phase supplies from two or three phases of the supply system results in voltage unbalance, and these two factors can lead to excessive propagation of negative-phase-sequence currents. Although it is possible to connect traction supplies to systems operating at MV or HV, the need to limit disturbances to other customers generally requires that the point of common coupling must be at the supply authority's HV distribution level.

To avoid load unbalance on the HV system it is usual to connect each substation across a different pair of phases on the HV side, as shown in Figure 12.15. Under normal conditions the HV/25 kV transformers feed the 25 kV overhead feeders in opposite directions as far as an intermediate switching station halfway between the substations. The arrangements are such that, for the total loss of any one HV/25 kV substation, due either to some local fault or a fault on the HV system, then trackside supplies are available from the substations on either side of the faulted substation. The average distance between substations is approximately 35 km on single-track lines and about 50 km on double-track lines.

From the load-current waveform it is possible to determine the magnitude of the individual harmonic currents and derive the total harmonic distortion at the point of common coupling. Experience suggests that the RMS value of the HV supply-voltage distortion can be around 2–2½% without causing undue interference to other loads. It should also be borne in mind that in taking traction supplies from only two phases of the HV supply the triple harmonics do not cancel out as with balanced 3-phase loads.

In areas with poor earthing conditions, the rail 25 kV overhead system is usually provided with booster transformers at approximately 3 km spacings in order to reduce induced voltages and noise interference in telecommunication

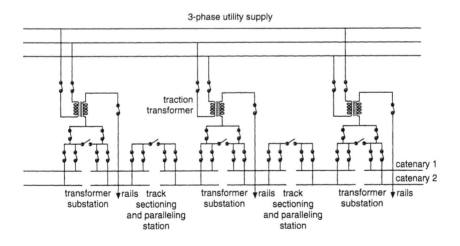

*Figure 12.15    Centre-fed AC railway traction feeding system with catenary-fault isolation arrangements*

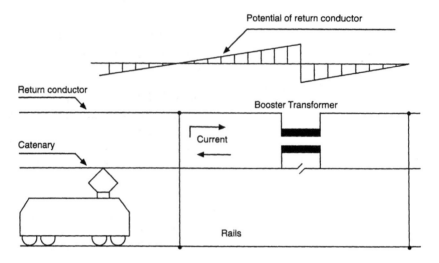

*Figure 12.16    Principle of booster transformer*

lines as well as the rail potential. These booster transformers are effectively current transformers with a 1:1 ratio and induce a voltage opposite to the voltage drop of the return rail conductor, offsetting the effect of the conductor reactance and forcing the load current to flow along the return conductor; see Figure 12.16.

With varying loadings on individual traction supply points the net unbalanced loading can result in excessive negative-phase-sequence currents circulating in the HV system. Consideration may then have to be given to phase balancing. Figure 12.17 shows typical connections for balancing a single-phase line–line load on a 3-phase system. It should be noted that the inductive arm of the balancer is connected between the phase not used for the railway load and the leading phase of the load, and that in practice it may be uneconomic to provide complete balancing.

## 12.6 Other loads

### 12.6.1 Welding equipment

Welding equipment usually draws a fluctuating current from the supply system, and therefore produces voltage fluctuations. Whilst these fluctuations are less erratic than those from arc furnaces, the limit of acceptability is the same, i.e. as shown in Figure 12.2. Consider a single-phase welder connected across two phases of a 3-phase system. A single-phase load on a 3-phase system involves a

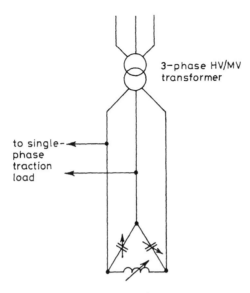

*Figure 12.17 Phase-balancing arrangements*

30° phase shift, and two of the phase–neutral voltages will experience voltage drops. The worst voltage drop is given by

$$\Delta V = I\{R_s \cos(\phi - 30°) + X_s \sin(\phi - 30°)\} \tag{12.8}$$

where $R_s$ and $X_s$ are the resistance and reactance at the welder supply terminals, respectively.

For a 400 V 3-phase system, and taking an average power factor for welding equipment of $\cos \phi = 0.3$, the approximation given by eqn. 12.9 can be used to determine the numerical value of the percentage voltage drop at the point of connection of the welder to the system, using the numeric values of the various quantities shown.

$$\Delta V\% \simeq (0.81R_s + 0.73X_s) \times S_w \tag{12.9}$$

where $\Delta V\%$ = percentage voltage drop caused by the welder, relative to the nominal 3-phase supply voltage

$R_s$ = resistance at welder supply terminals, $\Omega$/phase

$X_s$ = reactance at welder supply terminals, $\Omega$/phase

$S_w$ = apparent 'step kVA' of the welder, which can usually be taken as twice the nameplate rating

## 12.6.2 *Induction-heating equipment*

Induction-heating equipment basically falls into two categories, depending on the frequency of the power used for the heating effect. The induction-coil system

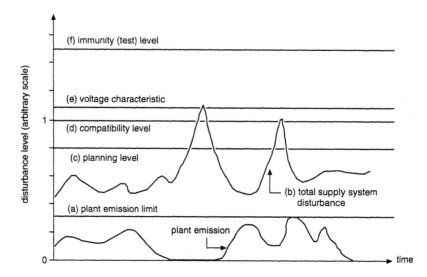

*Figure 12.18   Deterministic representation of the co-ordination of conducted disturbances (Courtesy UNIPEDE NORMCOMP 89)*

   *a*  Defined by standards or electricity supplier (maximum or 95% value)
   *b*  For a network location with a medium-high disturbance
   *c*  Defined by electricity supplier (time static, 95% value)
   *d*  Defined by standards (time and location statistic, 95% value)
   *e*  Defined by standards (time statistic, 95% value)
   *f*  Defined by standards or agreed between user and manufacturer

the user supply terminals is part of the problem. Relatively loose requirements for the voltage quality would result in high immunity levels for the equipment, which would ultimately make them very expensive. On the other hand, strict requirements would increase the utilities' investment costs, which would also affect the price of electricity.

The international standardisation work relating to electromagnetic compatibility and quality of electricity supply is carried out by the IEC. In European countries the standardisation work related to these issues is done at CENELEC. The general guidelines and requirements for legislation are established in directives issued by the Commission of the European Communities while the technical details (specific limits, test procedures, etc.) are presented in harmonised standards (ENs – European standards). Usually these harmonised standards are adopted directly in all the member countries of the European Communities as national standards. Two directives in particular have an important impact on the standardisation work related to the quality of electricity supply and EMC: the Product Liability Directive (85/374/EEC) and the EMC Directive (89/336/EEC). To clarify the application of the Product Liability Directive to electricity as a product CENELEC prepared the European standard EN 50160 'Voltage characteristics of electricity supplied by public distribution systems' published in January 1995. The standard specifies the

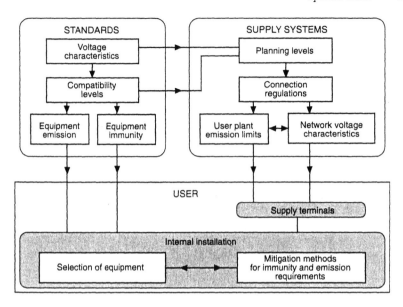

*Figure 12.19   Relation between standardisation, management of voltage characteristics, user
equipment and installation options
(Courtesy UNIPEDE NORMCOMP 89)*

characteristics of the voltage in public low-voltage and medium-voltage
electricity distribution systems. The requirements related to the safety of
electrical equipment are defined in the Low-Voltage Directive and associated
harmonised standards, which apply to equipment with nominal voltages of 50 to
1000 V AC and 75 to 1500 V DC. The Low-Voltage Directive is not directly
related to the EMC Directive, but is practical to test the conformity of the
product with both directives at the same time.

## 12.7.2  Voltage characteristics

Rapid voltage changes can be divided between single rapid voltage changes, for
example those caused by switching on a motor, which are described as a rapid
depression of the RMS value of the supplied voltage, and the voltage fluctuation
described as a series of consecutive voltage changes, caused for example by
electric arc furnaces or welding equipment.

Rapid voltage changes can be evaluated by measuring the change in voltage
RMS value compared with the nominal voltage, or by measuring and
calculating the flicker severity. Short-term severity $P_{st}$ is measured with a
specified equipment over a period of ten minutes and long-term severity $P_{lt}$ is
calculated from a sequence of $12P_{st}$ values over a two-hour interval, according to
the expression

$$P_{lt} = \sqrt[3]{\sum_{i=1}^{12} \frac{P_{sti}^3}{12}} \qquad (12.10)$$

According to EN 50160, for example, the voltage fluctuation should be less than or equal to $P_{lt} = 1$ for 95% of the time.

Supply-voltage dips are sudden reductions of the RMS voltage value to a magnitude between 90% and 1% of the nominal value followed by a voltage recovery, and the duration of a voltage dip is typically between 10 ms and 1 minute. The dips are mostly caused by single-phase or phase–phase faults and their effects depend on their depth and duration. A condition in which the voltage is lower than 1% of the declared voltage at the supply terminals is called supply interruption. Supply interruptions may be pre-arranged (e.g. due to scheduled maintenance or construction work in the distribution system) or accidental (caused mostly by external events, equipment failures or interference). Accidental interruptions may be classified as long (longer than 3 minutes) or short (less than 3 minutes) interruptions.

Temporary (power frequency) overvoltages are oscillatory overvoltages of relatively long duration, which are undamped or weakly damped, and are caused by switching operations or faults – for example, sudden loss of load, single-phase faults, nonlinearities. Transient overvoltages are short in duration (a few milliseconds or less) and usually highly damped and are mostly caused by lightning, switching, or the operation of fuses.

Harmonic voltage may be defined as a sinusoidal voltage with a frequency equal to an integer multiple of the fundamental frequency of the supply voltage. Correspondingly, interharmonic voltage is defined as a sinusoidal voltage with a frequency between the harmonics. Harmonic voltages can be evaluated individually or globally; individually by their relative amplitude $u_h$ related to the fundamental voltage $u_1$, where h is the order of the harmonic, and globally by the total harmonic distortion factor THD which is calculated as follows:

$$\text{TDH} = \sqrt{\sum_{h=2}^{40} u_h^2} \qquad (12.11)$$

The harmonics of the supply voltage are caused mainly by nonlinear loads and supplies. Harmonic currents flowing through the system impedance give rise to harmonic voltages. EN 50160 specifies the limits below which the 10-minute-average values of the harmonic voltages must remain for 95% of the time during each period of one week.

Mains signalling voltages may also degrade the quality of the voltage. The signalling voltages are superimposed on the supply voltage for the purpose of transmission of information in the public distribution system. Three types of signals are commonly used: ripple control signals (110–3000 Hz), power-line-carrier signals (3–148·5 kHz) and mains marking signals (transient signals).

High-frequency ( > 150 kHz) electromagnetic interference (EMI) is not included in EN 50160 but is regulated by the EMC standards. High-frequency EMI may be generated in medium-voltage distribution systems by electrical discharges occurring in insulators, disconnectors and other hardware. The highest interference levels are usually generated by gap- or corona-type discharges, and interference may also be generated by solid-state switching elements such as thyristors used in some power system equipment. Thus, some of the equipment used to increase the utilisation of power networks such as static VAR compensators (SVCs) may increase the overall EMI. The characteristics of EMI generated by these power electronic devices are completely different from those of electrical discharges. Rotating machines, saturated magnetic circuits (transformers, coils with magnetic cores), welding equipment etc. are also possible sources of EMI. The EMC Directive and associated standards set the requirements for EMI generated by power system equipment. CENELEC's standardisation programme on electromagnetic compatibility includes product family standards for high- and low-voltage switchgear, remote-control, protection and communication equipment and fuses. Product family standards define the specific limits and test procedures for the electromagnetic interference which equipment is allowed to produce (emission levels), and the interference levels which equipment must tolerate (immunity levels) and yet still maintain a predefined functional level.

## 12.7.3 Measurement of voltage characteristics

Measurements of voltage characteristics can be divided into three categories according to the purpose of the measurements: permanent monitoring (for example for verifying contractual obligations), temporary surveying (for example to check the performance of the supply system, or to check user complaints) and general investigations. Each category of measurements sets specific requirements for the instruments and the measurement methods. Portable power-quality instruments are most suitable for temporary surveying as they can be optimised for finding and solving intermittent power problems. Network power-quality instruments are used to gather long-term statistical data and to measure the general performance of a distribution system. They are optimised for data accumulation and communication, for example with a host computer.

In low-voltage supply systems the voltage to be measured can usually be connected directly to the instrument and the measurement is technically simple. In medium-voltage systems instrument transformers have to be used and the voltage-quality instrument then performs the measurement of various voltage characteristics by means of an analogue or digital data-acquisition technique. Modern power-quality instruments may be equipped with multiple micro-processors and a communication interface for data transmission to a host computer. They usually utilise digital signal processing (DSP) and may even include some level of expert advice to help the user of the instrument in interpreting the results.

*Chapter 13*
# Network voltage performance

## 13.1 General

The quality of electricity supply is considerably influenced by the quality of the voltage provided to customers, which can be affected in various ways. There may be long periods of variation from the normal voltage, sudden changes in voltage, rapid fluctuations, or unbalance of 3-phase voltages. In addition, other irregularities such as variations in frequency and the presence of non-linear system or load impedances will distort the voltage waveform, and transient spikes and surges may be propagated along circuits in a supply system. Some examples of these are shown in Figure 13.1*b*.

To avoid harmful effects to equipment belonging to the supply authority, or any customer, various forms of legislation and recommendations exist in

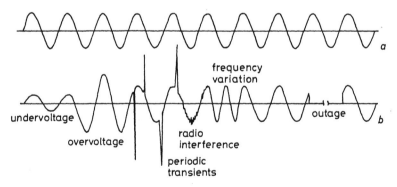

*Figure 13.1   Ideal voltage waveform and voltage variations*[1]

    *a* Ideal voltage waveform
    *b* Variations in voltage waveform

different countries to ensure that the level of the voltage supplied does not go outside prescribed tolerances. The characteristics of the supply voltage specified in voltage-quality standards usually describe its frequency, magnitude, waveform and the symmetry of the three phase voltages. Worldwide there is a relatively wide variation in accepted tolerances related to different voltage characteristics. Standards are constantly being developed in order to respond to technical, economic and political evolution.

Because some incidents affecting the supply voltage are random in time and location, some of the characteristics may be described in the standards with statistical parameters instead of specific limits. An important aspect in applying the standards is to look at where in the supply network, and when, the voltage characteristics are specified. The European standard *EN 50160*, for example, specifies the voltage characteristics at the customer's supply terminals under normal operation conditions. 'Supply terminals' is defined as the point of connection of the customer's installation to the public system.

*EN 50160* indicates that, in the member states of the European Communities, the range of variation of the 10 minute RMS values of the supply voltage (phase–neutral or phase–phase) is $V_n \pm 10\%$ for 95% of a week. For 4-wire 3-phase systems $V_n = 230$ V between phase and neutral. Strictly speaking, this means that for more than 8 hours a week there are no limits for the supply voltage value. There has also been some criticism that the voltage tolerance of $V_n \pm 10\%$ is too wide. Until the year 2003, the nominal voltage and the tolerances may differ from the values stated above according to the harmonised document *HD 472 S1*. During this transitional period, the countries having 220/380 V systems should bring the voltage to 230/400 V+6%/−10% and those countries having 240/415 V systems should bring the voltage to 230/400 V +10%/−6%.

The frequency of the supply system depends on the interaction between generators and load and the range of variation is smaller the higher the ratio between generation capacity and load. This means that it is more difficult for small isolated supply systems to maintain an accurate frequency than for systems with synchronous interconnection to adjacent systems. In the European Communities the nominal frequency of the supply voltage is 50 Hz. According to *EN 50160* 'the average value of the fundamental frequency measured over 10 s in distribution systems with synchronous connection to an interconnected system shall be within a range of 50 Hz ±1% during 95% of a week and 50 Hz +4%/−6% during 100% of a week'. Distribution systems with no synchronous connection to an interconnected system have wider tolerances of ±2% and ±15%, respectively. The frequency tolerances of *EN 50160* are also rather wide compared with the present situation in many member states.

In a series of studies on customer voltage variation, one UK electricity utility recorded the maximum and minimum voltages of every customer for each one hour period. From this information the mean values of these maximum and minimum voltages were plotted for all the customers, as shown in Figure 13.2. It can be seen that those customers whose voltage level was high had a small variation in their voltage. On the other hand, those customers who received a lower level of voltage experienced larger variations in the voltage throughout the 24 hour period. Often it is this varying low level of voltage which is a cause of annoyance to customers, even though the actual voltage received may be within the prescribed limits.

Where no customer equipment is directly connected to a network, which generally applies to those operating above 100 kV, there is no need for such a precise voltage level. Consequently the voltage levels are determined from considerations of power flows and system losses. The tapping range on step-down transformers from these networks will have to be sufficient to cope with these variations in voltage level, as well as with the internal voltage drop in the transformer and any compounding required for the secondary medium-voltage network.

All networks experience voltage drops on each circuit proportional to the loading, which is continually varying. Compensating equipment is therefore provided at suitable points on the various networks to offset the resultant variations in voltage. For example, automatic on-load tap changers on EHV/ HV and HV/MV transformers maintain the voltage at the HV and MV busbars, respectively, within acceptable operational limits. In some systems MV/LV transformers have tappings which can be selected off load to take account of voltage drops on the MV network, the MV/LV transformers and through the LV network.

Overall the various voltage-control equipments are operated in such a way that the voltage provided to MV and LV customers remains within the required limits, despite varying voltage drops due to changing loads and alterations in network configurations. Although a large variety of methods of compensating for voltage drop are available, such equipment increases the complexity of network operation and maintenance. Where the chosen MV level is high, e.g. 20 or 30 kV, voltage-drop problems are rare and usually the regulation provided by automatic tap changers on the HV/MV transformers is adequate.

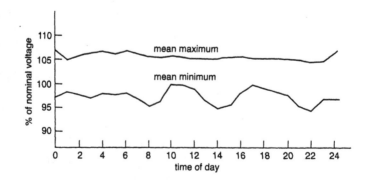

*Figure 13.2    Variation of customers' voltage over 24 h period*

## 13.2 Voltage regulation

Figure 13.3*a* diagrammatically represents part of a distribution network. For simplicity the voltage variation along one MV feeder only will be considered together with the effect of any distribution-transformer tapping, the distribution-transformer load voltage drop, as well as the voltage drop along the LV distributor and the individual service connection to a customer.

Figure 13.3*b* shows diagrammatically the relative voltage variation along the MV and LV systems when the networks are heavily loaded. The distribution-transformer tap has been set to give voltage boost to offset the MV feeder voltage drop, and to keep the LV voltage within acceptable limits. The manner in which the various voltage-control facilities are adjusted to ensure that customers receive a voltage within specified limits is covered in Section 13.7.

One problem is how the total voltage drop from the HV/MV substation to the furthest LV customer should be divided across the various elements of the system. In principle this can be solved by using cost functions for MV lines (see Figure 14.9), distribution substations and LV lines, and then scanning the network arrangement until the required total voltage drop, e.g. 12%, is achieved and the incremental costs, e.g. £/%, for the MV and LV lines are equal. The solution is dependent on load distributions and MV line lengths and on the unit cost values applied.

An example of such a study is shown in Figure 13.4 which gives the optimum voltage drop for a 20 kV overhead line as function of the feeder length. Distribution conditions, for example rural or urban, will influence the result. The distribution transformer load voltage drop may sometimes be critical and

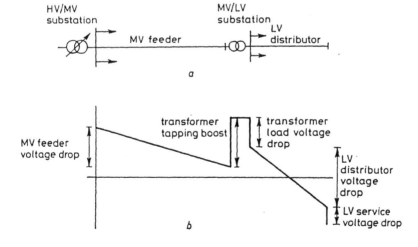

*Figure 13.3   Voltage regulation on MV and LV networks*

    *a* Simplified distribution network diagram
    *b* MV- and LV-network voltage variation

and is often termed line-drop compensation since it compensated for variation in the MV-feeder voltage drop.

Line-drop compensation, also known as voltage compounding, is applied to the voltage-regulating relay controlling the voltage of a source busbar, so that the busbar voltage is varied depending on the load supplied from that busbar. The compensation is achieved by injecting a current proportional to the transformer load current (derived from a CT) through an impedance $r_S$ and $x_S$ adjusted to model the network impedance. The resultant voltage drop is combined with a voltage proportional to the controlled busbar voltage (derived from a VT) to operate a voltage regulating relay. The voltage-regulating relay is maintained in balance by causing the supply-transformer tap changer to operate

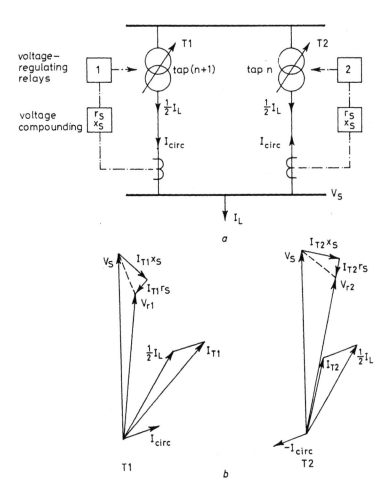

*Figure 13.6*    *Positive-reactance compounding*

  *a* Schematic
  *b* Phasor diagrams for T1 and T2

and change the busbar voltage up or down as necessary. The arrangement is shown schematically is Figure 13.5.

Figure 13.6*a* shows two transformers in parallel, each with a voltage-regulating relay, and compounding with a positive-reactance setting of $x_S$. Owing to equipment tolerance one transformer will tap-up first, say T1. The combination of load current $I_L$ and current circulating between the two transformers $I_{circ}$ causes the regulating relay voltages $V_{r1}$ and $V_{r2}$ to be as shown in Figure 13.6*b*. It will be seen that this arrangement is unstable. T1 regulating-relay voltage $V_{r1}$ is now reduced and tending towards another tap step-up, whereas T2 regulating-relay voltage $V_{r2}$ is increased and tending to tap-down. This process leads to tap divergence.

With the negative-reactance compounding, settings of $r_S$ and $-x_S$ are used. For the same condition as illustrated in Figure 13.6*a*, i.e. with current circulating owing to a difference in tap positions between T1 and T2, the resultant phasor diagrams are as shown in Figure 13.7. T2 regulating-relay voltage $V_{r2}$ is depressed and is thus tending towards a tap change to bring the transformers on to the same tap. However, T1 regulating-relay voltage $V_{r1}$ is increased and therefore has no tendency to tap up again.

Thus stable parallel operation can be achieved by the use of negative-reactance compounding, although changes in load power factor will alter the voltage rise obtained. For lagging power factor an improvement in power factor will cause an increase in voltage $V_S$ at the regulated busbar. In addition to providing stable parallel operation at one substation, negative-reactance compounding schemes permit the parallel operation of transformers at different substations on the lower-voltage networks.

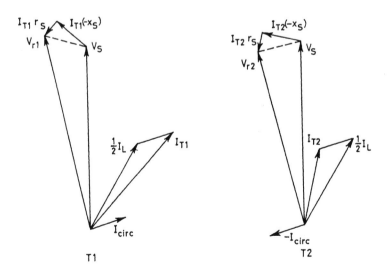

*Figure 13.7   Phasor relationships for negative-reactance compounding*

While automatic voltage control is applied at most EHV/HV and HV/MV transformer substations, the use of line-drop compensation (voltage compounding) is restricted to some HV/MV substations in certain supply organisations. In some cases facilities exist for adjustment of the line drop compensation setting by remote control.

## 13.5 Distribution-transformer tap settings

In MV/LV substations the distribution transformer can be provided with tapped MV windings which permit off-load adjustment of the transformer MV/LV ratio. The actual tapping used is determined by the need to maintain the voltage supplied to low-voltage customers within the required tolerances. Typically, the range of tappings is ±5% in 2½% or 5% steps either side of the nominal ratio. The tapping selected depends on the voltage drop along the MV feeder, the voltage drop through the distribution transformer, which is a function of the transformer loading, the maximum low-voltage-distributor voltage drop, and the automatic voltage control and line-drop compensation arrangements on the source HV/MV transformers.

Generally there are zones in a distribution network within which standard distribution transformers can all operate on the same tapping. Calculations to optimise the voltage conditions across the MV and LV networks are generally aimed at identifying and defining the boundaries of these zones. In some boundary areas of the network there may be an overlap of zones with a subsequent choice of tapping. If the loading on a particular distribution transformer does not bear the same relationship to transformer rating as the other transformers in a 'tap zone', or its associated low-voltage-network voltage drop varies considerably from the average value within the zone, it may be necessary to utilise a different tap from that used for the particular tap zone. Usually one tap higher or lower is adequate. There is evidence that, where voltage compounding is in use on the MV busbar of the associated infeed HV/MV transformer substation, most MV and LV networks operate satisfactorily with a range of 5% being used for the distribution-transformer tappings.

The major difficulty in resolving a distribution-network voltage problem is the co-ordination of the various voltage-control facilities, given that much of the information required is not readily available or is of unknown accuracy. However, in practice the tap setting of individual distribution transformers supplied from one HV/MV substation is rarely critical. Experience suggests that typically only 2 or 3% of distribution-transformer tappings may require special investigation, and the incidence of justified customer voltage complaints would indicate that the use of one tapping for a given zone of the network, as discussed earlier, is a sensible engineering approach.

Opinions vary concerning the usefulness of tappings on distribution transformers. If the MV network open points have to be altered owing to network faults, the tappings in use could result in uneven LV-network voltages.

Whilst under adverse conditions these may be outside the prescribed limits, in practice customers are rarely inconvenienced, since, in general, any one distribution-transformer tap setting is not critical to the local LV-network voltage under most system operating conditions. In distribution systems where medium-voltage drops are relatively low, and the voltage elasticity on the LV networks is also small, there may be little benefit in using MV/LV transformers with taps.

## 13.6 Regulators and capacitors

If reinforcement of a network were required because of excessive voltage drop(s), any such reinforcement could be deferred if the voltage drop could be sufficiently reduced by some means, subject to economic as well as the technical considerations. Various devices can be installed to reduce the voltage drop, or voltage range, experienced at critical points on MV networks. Secondary considerations are that, by maintaining the correct voltage tolerances along the feeder, this could also possibly permit the use of distribution transformers without taps.

One method of maintaining the voltage along the feeder is the use of a voltage regulator. Figure 13.8 provides a single-phase representation of a moving-coil regulator. This type of regulator contains two coils A and B of similar impedance connected in series opposition. Two secondary windings *a* and *b* are associated with the main windings, shunted across the input and mounted on the same magnetic core as the main coils. Another coil m is short-circuited upon itself, and can be moved along the magnetic core by hand or motor, the power requirements for movement of the coil being fairly small. In the example shown, winding *a* has 21 times the number of turns of winding A, and winding b has nine times the number of turns of winding B.

To raise the voltage from 0·9 to 1·0 p.u., coil m would be located as shown in Figure 13.8. The impedance of winding *a* would be small owing to the effect of the short-circuited winding *m*, and that of *b* would be relatively large so that effectively the voltage across the regulator appears across winding *b*. With 0·9 p.u. in winding *b* and a 9:1 ratio, the voltage induced in winding B is 0·1 p.u. With zero voltage induced in winding A the output voltage is thus 0·1 p.u.

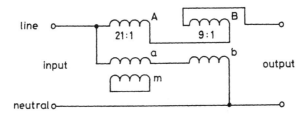

*Figure 13.8 Schematic of moving-coil voltage regulator*

values. The customer voltage is thus the summation of the MV-busbar voltage, and the voltage gains and drops through the network, as shown at the bottom line of Table 13.1*a* and *b*.

Given that the maximum variation in voltage permitted is between 95% and 105%, then, from Table 13.1*a*, at maximum load customer D has a supply voltage below tolerance, and from Table 13.1*b* all customers are above voltage limits at time of minimum load. Table 13.2 summarises the situation at maximum and minimum loadings, and then details in the left-hand row the successive actions to bring the customers' voltages within limits.

With an MV-source busbar voltage of 103% at all times, customer voltages vary from 92·5% (D at maximum load) to 107·2% (A at minimum load). By adjusting the voltage compounding on the HV/MV source transformer to 105·5% at time of maximum load and 100·5% at time of minimum load, this variation is reduced to 95·0–106·0%. However, customer A is still outside limits at maximum load (106·0%) and very close to the required value of 105% at minimum load.

Adjusting the tapping on distribution transformer X to reduce the voltage boost from 5% to 2·5%, as shown in lines 5 and 6, results in supplies to all

*Table 13.1   Voltage-drop example*

(a)  Maximum load conditions: percentage voltage values

| Customer | A | B | C | D |
|---|---|---|---|---|
| MV-busbar voltage | 103 | 103 | 103 | 103 |
| MV-feeder voltage drop to | | | | |
|    distribution transformer | −0·5 | −0·5 | −6 | −6 |
| Transformer-ratio change | +5 | +5 | +5 | +5 |
| Transformer-load voltage drop | −2 | −2 | −2 | −2 |
| LV-network voltage drop | −0·5 | −6 | −0·5 | −6 |
| LV-service voltage drop | −1·5 | −1·5 | −1·5 | −1·5 |
| Customer voltage | 103·5 | 98 | 98 | 92·5 |

(b)  Minimum load conditions: percentage voltage values

| Customer | A | B | C | D |
|---|---|---|---|---|
| MV-busbar voltage | 103 | 103 | 103 | 103 |
| MV-feeder voltage drop to | | | | |
|    distribution transformer | −0·1 | −0·1 | −1·0 | −1·0 |
| Transformer-ratio change | +5 | +5 | +5 | +5 |
| Transformer-load voltage drop | −0·35 | −0·35 | −0·35 | −0·35 |
| LV-network voltage drop | −0·1 | −1 | 0·1 | −1 |
| LV-service voltage drop | −0·25 | −0·25 | −0·25 | −0·25 |
| Customer voltage | 107·2 | 106·3 | 106·3 | 105·4 |

*Table 13·2 Action to improve customer voltages*

| Action | Customer voltage (% of nominal) | | | |
|---|---|---|---|---|
| | A | B | C | D |
| *Situation as specified* | | | | |
| 1 Maximum load | 103·5 | 98·0 | 98·0 | 92·5 |
| 2 Minimum load | 107·2 | 106·3 | 106·3 | 105·4 |
| *Voltage compounding* | | | | |
| 3 105·5% at maximum load | 106·0 | 100·5 | 100·5 | 95·0 |
| 4 100·5% at minimum load | 104·7 | 103·8 | 103·8 | 102·9 |
| *Tap setting of transformer X changed from 5% to 2·5%* | | | | |
| 5 Maximum load | 103·5 | 98·0 | | |
| 6 Minimum load | 102·3 | 101·3 | | |
| *Tap setting on transformer Y unchanged* | | | | |
| 7 Maximum load | | | 100·5 | 95·0 |
| 8 Minimum load | | | 103·8 | 102·9 |

customers being within the required voltage range of 95–105%. The variation in supply voltage of each customer has also been considerably reduced from the initial situation. For customer A the variation has dropped from 3·7% to 1·3%, for customers B and C from 8·1% to 3·3%, and for customer D from 12·9% to 7·9%.

Thus by suitable values of source busbar voltage, voltage compounding on the HV/MV transformer(s) supplying the MV networks, and adjusting distribution-transformer tappings, acceptable voltage levels can be maintained for the LV customers across the range of system loading. Interactive computer programs can be used to improve the performance of these voltage-control facilities, with the engineer having the facility to fix parameters at individual locations to meet specific network requirements.

## 13.8 Voltage unbalance

Voltage unbalance is usually expressed in terms of a voltage unbalance factor which is defined as the ratio of the negative-phase-sequence voltage component to the positive-phase-sequence voltage component.

Voltage unbalance is caused by unbalanced phase impedances, or unbalanced loads, or a combination of both. Unequal phase impedances arise on horizontal- or vertical-formation lines due to the asymmetrical conductor spacing, so that the centre phase presents approximately 6–7% lower impedance than the outer two phases.

Unbalanced load conditions can arise on MV rural systems where single-phase distribution transformers and spur-line supplies are tapped off the 3-phase network and the loads of these tappings are not balanced across the three phases. The load unbalance on individual LV distributors can be considerable and vary with time of day but, with three or four LV distributors from each MV/LV substation, the overall effect will be less pronounced at the substation. On an MV feeder supplying a number of MV/LV substations, the overall unbalance, due to LV unbalance, will generally not be very significant.

The negative-sequence impedance of a motor is much lower than the positive-sequence impedance. With voltage unbalance any resultant high negative-sequence currents in motors can lead to high temperature rises. Whilst some motors are equipped with protective devices which trip out the motor when temperatures becomes too high, there have been cases of motors overheating which have been attributed to excessive voltage unbalance. In addition, negative-sequence current causes reverse torque which tends to retard the motor. International experience indicates that some voltage unbalance is acceptable provided that this is under 2%, and preferably below 1.5%.

Most voltage-compounding equipment operates from the load current in one phase and the phase voltage in the other two phases. With voltage unbalance the equipment will incorrectly adjust the transformer tap position, assuming that these values represent a balanced system. Similar problems may occur with protection equipment.

Two types of winding connections are used in 3-phase distribution transformers, either the delta/wye (Dy) or Wye/zigzag (Yz) connection. The connection of the distribution transformer must be chosen so that any unbalanced secondary loading causes minimum distortion of the MV-phase voltages.

Any reduction in voltage unbalance will reduce network voltage drop and will also reduce system losses. In general, it is the load unbalance which is the major cause of voltage unbalance. The connection of single-phase loads to 3-phase supplies also results in higher losses.

## 13.9 Constraints affecting network-voltage performance

Apart from the limitations of the tap steps at the primary substation or on the distribution transformers, and the problems of voltage unbalance, other factors influence the ability to provide optimum voltages to consumers. Two main problems are the variation in loadings across the distribution network and the arrangements necessary to cover outage conditions. Loading patterns on the various distribution transformers will be different, so that the voltage drop on individual feeders will be different and vary with each other with time. The degree of line-drop compensation that can be applied may be limited. For example, if the amount of compensation is related to the general loading on the HV/MV substation, the compensation could be incorrect for any feeder supplying loads which have significantly different daily load curves.

Under fault or maintenance conditions on an MV feeder it is necessary to supply the MV/LV distribution transformers on the remaining healthy section of the MV feeder. Figure 13.11a shows normal operating conditions on a typical distribution network. For a fault or maintenance outage between points d and e on the lower MV feeder, this section of feeder would be isolated by opening the disconnectors at points d and e as in Figure 13.11b. Closing up at the normally open point at c restores supplies to the distribution transformers between c and e. Under this situation the healthy MV feeder a–b providing the back-up supply will be more heavily loaded, with a consequent increase in the MV feeder voltage drop. The distribution transformers along its length will thus be subjected to a reduced voltage with a consequent lowering of the voltage of the association LV networks.

In this situation, those distribution transformers receiving standby supply, e.g. those between points c and e, will similarly receive a lower voltage on the MV side. The resultant effect on the LV networks supplied from these latter transformers will be furthered worsened by the fact that, in many cases, the power flow down the remaining section of the faulted MV feeder will now be in the opposite direction to normal flows. If the distribution transformers have preset tappings, this reversal of power flow could, in certain adverse conditions, result in the voltage received by some customers being outside normal tolerances. In urban situations the low-voltage supplies from the distribution transformers on the disconnected section of MV feeder will generally be redistributed and supplied from other transformers outside the disconnected section of the MV feeder, as indicated in Figure 13.11b.

If the fault is such that a number of distribution transformers have to be switched to an adjacent MV substation, as in Figure 13.11c, the loading on that substation will be increased. The degree of additional loading will depend on the number and the loading of the distribution transformers transferred to the second MV substation. With line-drop compensation in operation at the second substation, the MV busbar voltage will be raised. There may thus be a need to limit the range of line-drop compensation (LDC) to avoid unacceptable voltages under such outages.

## 13.10 Customer-voltage fluctuations

There are a number of factors which cause irregularities and fluctuations in the voltage supplies to customers' installations. Some of these problems are caused by equipment within a customer's own installation, e.g. the opening of switches or contactors, and appliances using thyristors or triacs. These may be sufficiently attenuated so as not to cause annoyance to adjacent customers, although they may affect or even disrupt other equipment within the customers' premises.

Irregularities in the supply voltage can cause various problems depending on the nature of the disturbance. The quality of the voltage depends not only on the degree to which these irregularities are propagated throughout the various networks, but also on the sensitivity of electrical appliances to such irregularities. It is therefore necessary to consider the effects of a sudden step change in voltage,

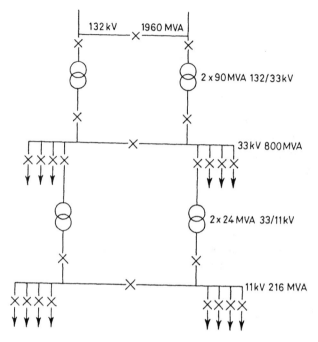

*Figure 13.12   Network for impedance/frequency study*

increased system frequency, they may become overloaded. It is also possible to obtain a resonance condition in capacitor banks, and if the resonant frequency is close to a harmonic present in the system, large over-voltages are possible.

Figure 13.13 gives the impedance/frequency characteristic for the 11 kV busbar of the network shown in Figure 13.12, measured over the frequency range 50–1000 Hz. The characteristic has the typical 'hill and valley' curve, the

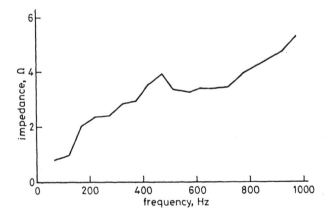

*Figure 13.13   Impedance/frequency characteristic for 11 kV busbar in Figure 13.12*

peaks being due to parallel resonance occurring within the system and the troughs being caused by series resonance.

In order to reduce harmonics, harmonic filters can be connected into the appropriate MV or LV circuit. Where there are several harmonic-generating loads, each fed by a distribution transformer, it is often more economical to eliminate the harmonics by installing filters at the MV busbar, rather than to have separate filters on the LV side of each transformer. Harmonic-filter arrangements, and their impedance/frequency characteristics, are shown in Figure 12.11.

Reference should be made to Section 12.7.2 regarding the calculation of total harmonic distortion (THD).

## 13.12 Bibliography

BARNES, R.: 'Harmonics in power systems', *Power Eng. J.*, 1989, **3**, (1), pp. 11–15

BARAN, M. E., and WU, F. F.: 'Optimal sizing of capacitors placed on a radial distribution system', *IEEE Trans.*, 1989, **PWRD-2**, pp. 735–743

BERGEAL, J., and MOLLER, L.: 'Influence of load characteristics on the propagation of disturbances'. IEE Conf. Publ. 197, International Conference on Electricity Distribution – CIRED 1981, pp. 61–65

BRADLEY, D. A., MORFEE, P. J., and WILSON, L. A.: 'The New Zealand harmonic legislation', *IEE Proc. B*, 1985, **132**, pp. 177–184

BRAUNER, G.: 'Calculation of the harmonics distribution in networks'. AEG Telefunken Prog., 1982, (2), pp. 41–46

BURTON, D. P., HEFFERMAN, D., and RICHMOND, E.: 'Optimiser; a microcomputer-based current controller to improve voltage levels in rural locations', *IEE Proc. B*, 1986, **127**, pp. 46–55

BYARS, M.: 'Voltage imbalance survey at a number of 132/33 kV bulk supply points' in 'Sources and effects of power system disturbances'. IEE Conf. Publ. 210, 1982, pp. 77–81

'Characteristics of the low voltage electricity supply' in 'l'Economie electrique'. UNIPEDE report 92, May 1981

CLEMMENSEN, J., and FERRARO, R.: 'The emerging problem of electric power quality', *Public Util. Fortn.*, 28 Nov. 1985

COMMELLINI, E., and TONON, R.: 'Voltage regulation in MV-LV distribution networks'. 7th International Conference on Electricity Distribution – CIRED 1983, AIM, Liege, Paper a17

DE VRE, R.: 'Harmonic distortion produced in supply networks by television receivers and light dimmers'. IEE Conf. Publ. 210, 1982, Electron. Pwr., Sources and effects of power system disturbances, pp. 123–128

DUGGAN, R., and MORRISON, R. E.: 'Prediction of harmonic distortion when a non-linear load is connected to an already distorted supply', *IEE Proc. C*, 1993, **140**, pp. 161–166

FARRELL, F. J.: 'Voltage criteria on a rural distribution system'. CIRED 1981, pp. 355–359

FOOSNAES, J. A., SAND, K., and SOLVANG, E.: 'Positioning of shunt capacitors in rural distribution networks'. 7th International Conference on Electricity Distribution – CIRED 1983, AIM Liege, Paper e03

GRAINGER, J. J., and CIVANLAR, S.: 'Volt/Var control on distribution systems with lateral branches using shunt capacitors and voltage regulators: Part I–III', *IEEE Trans.*, 1985, **PAS–11**, pp. 3278–3297

GRETSCH, R., and KROST, G.: 'Measurement and calculation of harmonic impedances of loads, LV and MV networks'. 7th International Conference on Electricity Distribution – CIRED 1983, AIM, Liege, Paper c02

HARRISON, D. H.: 'Voltage unbalance on rural systems'. *Power Engng. J.*, 1987, pp. 149–153

HOLMES, E. J., and MOFFATT, A. M.: 'Parallelogram theory for solution of distribution network voltage problems by computer', *IEE Proc. C*, 1983, **130**, pp. 153–164

'IEC standard voltages', IEC Publication 38, 1983

JANSSEN, JOH.: 'Quality monitoring in the Netherlands electricity distribution grid'. Third International Conference on Power Quality: End-Use Applications and Perspectives (PQA 94), 1994, Amsterdam, paper E-1.02

KEARSLEY, S. J.: 'A low cost harmonic analyser for the distribution network in Sources and effects of power system disturbances'. IEE Conf. Publ. 210, 1982, Sources and effects of power system disturbances, pp. 172–176

KIDD, W. L., and DUKE, K. J.: 'Harmonic voltage distortion and harmonic currents in the British distribution network, their effects and limitations'. IEE Conf. Publ. 110, 1974, Sources and effects of power system disturbances, pp. 228–233

LAKERVI, E.: 'Voltage drop limitation as a cost factor in electricity network design'. *Sähkö – Electricity and Electronics.* 1984, (6), pp. 14–17

LEHTONEN, M: 'Probabilistic methods for assessing harmonic power losses in electricity distribution networks'. IEE Conf. Publ. 338, Third International Conference on Probabilistic Methods Applied to Electric Power Systems – PMAPD 91, 1991, pp. 37–42

MORRELL, J. E., KENDALL, P. G., and CORNFIELD, G.: 'The application of the U1E flicker-meter to controlling the level of lighting flicker'. 9th International Conference on Electricity Distribution – CIRED 1987, AIM, Liege, Paper c 06

MÖRSKY, J., and SEPPÄLÄ, J.: 'Electricity supply quality in network planning'. 9th International Conference on Electricity Distribution – CIRED 1987, AIM, Liege, Paper a 01

PAPADOPOULOS, M., POLYSOS, P., FAKAROS, A., and MARIOOLIS, J.: 'Selection of the optimum size and location of capacitor banks on medium voltage networks'. 7th International Conference on Electricity Distribution – CIRED 1983, AIM, Liege, Paper e 02

PRICE, J. J., and GRIMWADE, D.: 'Aspects influencing the design of equipment to alleviate harmonic distortion in electrical supply systems. IEE Conf. Publ. 210, 1982, Sources and effects of power system disturbances, pp. 154–159

SACHER, Y., LE GAL, G., and BATTAGLIA, B.: 'Voltage control and regulation policy in distribution networks', *Rev. Gén. Elect.* July 1980, pp. 49–63

SELJESETH, H., PLEYM, A., and SAND, K.: 'The Norwegian power quality project'. Third International Conference on Power Quality: End-Use Applications and Perspectives (PQA 94), 1994, Amsterdam, paper E-1.037

TOBIN, E.: 'Selection of transformers for use in distribution networks', *J. IEE*, 1949, **96** Pt. 2, p. 485

'Voltage characteristics of electricity supplied by public distribution systems', European Standard En 50160

YACAMINI, R.: 'Power system harmonics', *Power Eng. J.*, 1994, **8**, (4), pp. 193–198

*Chapter 14*

# Computer-based planning

## 14.1 General

In distribution-network design a large amount of data is required, e.g. information on the present networks, design objectives, cost parameters and possible ways of reinforcement. Complicated calculations are necessary in some cases to optimise network configurations. The use of computers makes it possible to carry out sophisticated network-design calculations. The main aim of using computers here is to improve the quality of routine design.

Common tasks for computer-aided network design are obtaining quantitative information on the status of networks or determining the most suitable future network configuration and the optimum circuit ratings. Computer programs can also act as an efficient tool for long-term planning and the study of more complex aspects such as network reliability.

In practical network design the computer serves as a tool for the designer. Referring to Figure 14.1, many policy decisions such as voltage steps, unit sizes etc. must be fixed beforehand. Also clearly defined data on the existing network configuration and its components, and on feasible solutions, are necessary. The central block, 'modelling of system structure at planning horizon' includes the computer hardware and software, plus the designer's actions. The example in Section 14.6 will illustrate further the planning procedure.

It is usual for computers also to be used for the basic investigations shown in the right-hand column of Figure 14.1, but these are carried out separately from the modelling studies. The network-design programs are preferably integrated with other functions of a utility. It is thus desirable to create a common data bank to cover a large proportion of the information requirements of a utility, and to ensure that it is available to all sections of the utility.

has not been included. Most applications now used in utilities are still alphanumeric and, although the more recent ones are interactive, the lack of graphics limits their flexibility. This type of network information system is often called a facilities management (FM) system.

*Automated mapping systems*

Computer-aided design (CAD) systems also spread rapidly in the 1970s to urban electricity utilities. In spite of the work 'design' in the title, CAD systems are mainly oriented towards drawing. They provide a flexible tool for producing and handling accurate geographic maps for underground cables. Location information of network components can often be transferred from portable measuring equipment in the field (tachymeter) to the system. This can then be completed from the terminal to include information about the excavation, cable and additional map symbols — see Figure 14.2.

National or municipal ordnance survey authorities often produce various maps in digital form with equivalent CAD systems, and so background maps are quite easily available. With CAD systems views with different scales which are independent of the traditional map sheets can be obtained. Conventionally these systems are separate from network calculation applications. CAD systems which are used for producing digitised maps are often called automated mapping (AM) systems.

*Graphic network information system*

A disadvantage of FM systems is the poor clarity of the result and the limited possibilities for interactive design. AM systems require an additional and partly parallel database for line and plant information. Thus a strong demand was introduced for integrating FM and AM systems or for developing new network information systems with such features.

Typically, such a system has an interface showing the geographic background and network maps. The display often also includes several windows where alphanumeric information on different objects is displayed. Graphic and alphanumeric data can be updated at the same time and the attributes of different objects can be found by pointing to the object on the screen with the cursor. In addition, these objects may be linked to scanned images such as photographs or drawings. The characteristic of the objects may be different in different views having, for example, a different scale. Numerical results can also be displayed linked to the network map, and different colours in line sections or nodes can be used to indicate, e.g. levels of fault currents or voltage drops.

These systems are often called AM/FM systems or AM/FM–GIS systems (geographic information system).

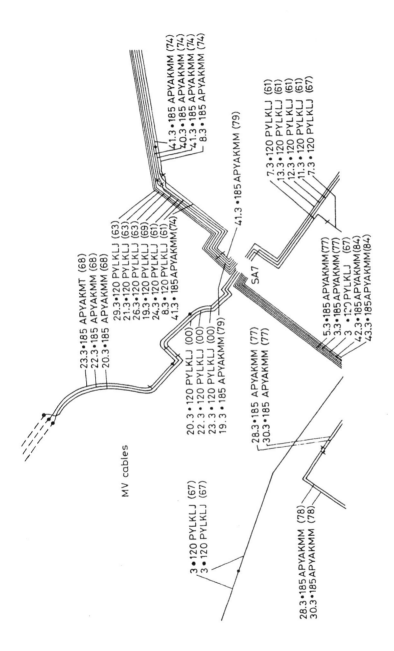

*Figure 14.2    Network map plotted by computer (courtesy Helsinki Energy Board)*

### 14.2.3 Information system components

There are many differences in the large number of commercial network information systems which are now available. All of them, however, can be illustrated by Figure 14.3 showing the general principle of database-oriented systems.

A database system consists of a database, a database management system (DBMS) and application programs, and the system also includes the necessary hardware such as the computers, display and printer units plus networks and system software such as the operating system and drivers.

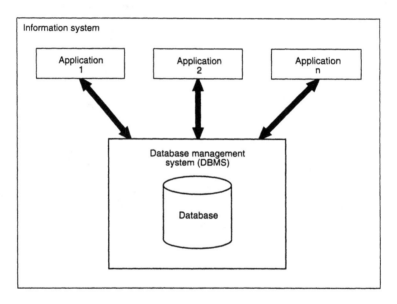

*Figure 14.3   Components of an information system*

### Database

For a utility the database is the most valuable part of the system, since it is the core of the stored and linked data. Physically the database consists of files usually stored on the hard disk. When the system is established or revised the data needed by the user and applications are modelled in a systematic manner, usually referred to as the creation of data models.

Data are stored or updated in the database only once and the same data are available for various applications and users. This is a very valuable feature because it prevents the existence of different parallel updated data. As shown in Figure 14.3, the stored data and application programs are separate from each other. A common supervision mechanism (DBMS) is applied for searching and updating of data.

*Database management system*

The database software is a general program which is used for managing and handling the data stored on disk. At present the most important class of database software is a relational database. These databases consist of tables and each object in a data model, for example a distribution substation or circuit breaker, can form a table. Links between tables are formed during enquiries. In this way each piece of information is stored in one location only and any set of related data can easily be retrieved simultaneously. The user can make enquiries to the database flexibly by using the Structured Query Language SQL.

Relational databases are flexible as regards revision and developments to the network information system, and are also well standardised which supports data exchange and linking between systems.

*Application programs*

When using a database system it is possible to apply the same interface program for many applications. This makes the development of applications more efficient. For the user it is convenient when all applications have similar interfaces. When application programs are separate from the data, new applications can be added without necessarily affecting the configuration of the database.

## 14.2.4 Network databases

Network information systems are often large and complicated and, unlike other data systems in a utility, their essential features are their capability for planning and handling graphic information.

*Contents of database*

For a utility most of the overall costs related to its network information system are concerned with the establishment and upkeep of its data. The main groups of data are network and plant data, data for day-to-day operation, work information and customer data. They include, for example, data related to the following items:

- construction, location and maintenance of LV, MV and street-lighting networks
- topology and connections between networks and substations
- faults and measurements
- consumptions and loads including load curves for customer groups
- customers and delivery points
- jobs in progress, labour costs and work results
- switching programmes, switching-state changes and measured data.

- busbar configuration
- data on auxiliary systems (telecontrol, protection, compensation etc.)
- construction data
- inspection data
- supporting data for plotting the main diagrams of the substation

## Transformer data

For each HV/MV and MV/LV transformer the following data are included in the database:

- location in the network
- manufacturer's serial number and year
- maintenance year
- exact technical specifications (power ratings, impedances, dimensions etc.)

## Switchgear data

The node codes in the above network and substation data provide a link to any switchgear records. Here the following data are specified for each device (disconnecter, recloser etc.):

- location in the network
- identification symbol for the device (name)
- type code
- voltage and current ratings
- breaking and making current ratings
- relay settings plus VT and CT ratios
- manufacturing year
- maintenance year
- statistical values (number of operations etc.)
- status of the equipment (closed/open)

## Other data

The technical values of conductors, data on all earthing arrangements, on LV cable boxes, and economic data for network design are also stored in the database.

## Load data

The loads of the existing customers required for network calculations are transferred from the customer database to the network database. Normally the transferred data include the code for each customer, the annual units and the codes for customer group and tariff. The codes used in the customer database for each customer must be compatible with the network node codes, so that it is possible to connect the loads to the right network nodes.

## 14.2.5 Database management

### Separate databases

AM/FM systems usually have two separate databases: a graphic database and an alphanumeric database. These are linked with each other. When graphic and attribute data of objects are revised the changes are made to both databases (Figure 14.6). Graphic information can be represented either in vector or raster mode (bit map). In the vector mode, for example, a line section can be presented when the co-ordinates of the ends of the section and the line type are known. In the raster mode a figure consists of pixels and their value gives the colour of that area element. Vector data are usually formed by digitising network maps with digitising tables, while raster data are produced with scanners which form a digitised file. In practice after automated scanning some manual adjustments are necessary.

As mentioned earlier, relational databases are popular for storing alphanumeric data. Several suppliers of CAD programs offer interface programs to most commercial relational databases like Oracle and Ingres.

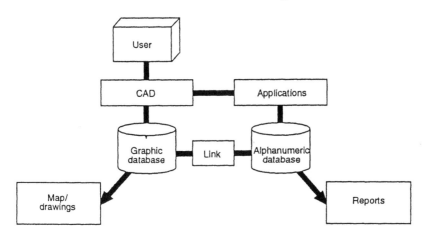

*Figure 14.6    Separate databases for graphic and alphanumeric data*

### A common database

In many of the recently developed network information systems the concept of one common database is applied. Only one relational database is used both for graphic and alphanumerical data. Figure 14.7 shows the main basis of such a system. The database does not usually include, for example, the 1:2000 network map pictures as these are formed outside the database whenever required.

From the utility's point of view a system with a common relational database is useful. Most utilities have the knowledge for operating and connecting new interfaces to a relational database and are thus also able to manage and develop

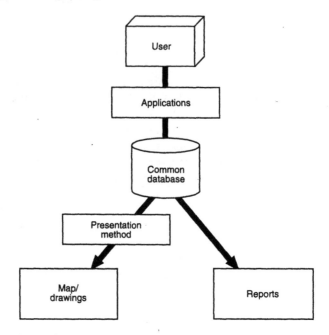

*Figure 14.7   A common general database for graphic and alphanumeric data*

the above system. The long response time faced earlier with relational databases has disappeared since the performance of computers and the management of the virtual memory have improved. When a large number of application programs have been implemented, even for a smaller utility the database may include hundreds of tables. The size of background map files outside the database may be up to several gigabytes.

## 14.2.6 Application programs

A modern network information system can include a large variety of application programs which support the management and development of the electricity distribution system and utilise the same database and a graphic interface.

### General applications

- maintenance and management of alphanumeric and graphic data
- printing and plotting of maps, schemes, drawings, pictures etc. related to existing or planned networks
- summaries and reports of equipment, annual performances of employees, etc.

### General planning

- management of area load forecasts
- generation of horizon-year target networks

- production of annual investment plans
- management of the capacity of the network and setting the development targets
- calculation of various parameters and summaries for the long-term development of the distribution system (e.g. degrees of utilisation of transmission capacities, losses, safety regulations, reliability).

*Planning and design*

- MV and LV planning (e.g. technical and economical comparison of alternative developments)
- production of documents for network tasks (e.g. basic work data, maps, estimated costs, construction, equipment)
- management of contracts (e.g. land use for lines and substations)
- management of construction standards.

*Construction*

- opening and co-ordination of tasks
- resource planning and management
- monitoring and reporting.

*Maintenance*

- management of inspection data
- analysis of maintenance and equipment data
- management of historical maintenance data
- management of the inspection and maintenance work programme.

*Operation of the MV network*

- operation planning (technical and economical)
- monitoring electricity delivered
- monitoring the switching state
- switching planning
- management of faults (e.g. supporting the identification of the fault location, restoration, collecting statistical data)
- trouble call activities.

An AM/FM-GIS system offers an effective and rapidly developing environment which can be used on more and more new applications and which effectively supports other graphic-related functions of the utility. In particular, the versatile capability for utilising geographic, network-map and substation-scheme information offers many possibilities for design, operation and customer services.

interruptions/customer/year. For planning purposes it is convenient to measure the reliability of the network with monetary values. The outage costs are thus calculated applying the value of non-delivered energy. Its value is based on customers' opinion of the inconvenience caused by the interruptions. Polling studies have been made about the value of non-delivered energy in various countries (see Section 4.2).

In the network monitoring function of the network information system the load flow and fault currents are calculated for the whole network of the utility. This is usually carried out once or twice a year. The results of monitoring calculations are given as summary sheets indicating, for example, electrically weak segments of the network. Based on these results, further studies are initiated in order to find out suitable ways to reinforce the network.

In the expansion planning of distribution networks the network-calculation modules have two functions:

- to provide the data needed for the calculation of the costs (e.g. power losses for the costs of losses); and
- to provide the data needed for checking that all the technical constraints have been achieved.

In an interactive planning procedure a possible future network is first designed so that the technical constraints are met. After that, the costs are calculated with the plan with the lowest costs being the preferred option. Sophisticated mathematical methods for network planning are discussed in Section 14.4.

The calculation modules can also assist in various operational tasks. If the network information system has a connection to the SCADA system, for example, on-line load-flow studies can be carried out. In an on-line study the current state of the system is simulated so that the calculations made with the aid of load curves are corrected using available measurements. The results can be used, for example, in estimating the effects of switching actions. The calculations needed for these tasks are only possible with the aid of load curves defined for the whole year.

Another important benefit of the connection between the network information and SCADA systems is the real-time information received about the states of switches so that the calculations can always be made using the actual switching configuration.

## 14.4 Mathematical methods for network planning

In addition to programs for network calculations, a modern network information system often includes modules for distribution network planning. For this purpose interactive design programs may be used, where the main tasks of the program are the calculation of the costs and checking the technical constraints. The design of different possible plans and the comparison of costs is made by the planner. However, in order to reduce the time required for planning and producing economically better plans, efficient optimisation models are needed.

During the last three decades several distribution planning models have been proposed. Very few of them have been successfully applied to practical planning tasks which illustrates well the complexity of the planning and design tasks faced in utilities.

The network planning task can be formulated as an optimisation problem where the object is to minimise the costs while taking into account the technical constraints. The objective function contains not only the investment costs, but usually also

- costs of losses
- costs of outages, and
- operational and maintenance costs.

A part of the constraints is always related to the structure of the network, e.g.

- all loads should be connected, and
- the network should be radial.

The constraints are also used for verifying that the solution satisfies

- safety regulations
- thermal capacities of the network components
- voltage drop limits, and
- operational requirements for the protection.

The planning model can be divided into four categories:

- static load/subsystem
- static load/total system
- dynamic load/subsystem
- dynamic load/total system.

In the static-load models (single-stage planning) the growth of loads is not taken into account, while in the dynamic-load models the loads are time varying and the planning process is viewed in several stages. In subsystem models only a portion of the distribution system is considered (usually either feeders or substations). The total-system approach considers the whole system, i.e. both the substations and the feeders. An exact mathematical model of the planning problem would lead to an extremely complicated model and therefore several approximations are usually made. The most common approximation relates to the cost function describing the total costs of a component as a function of the power flow. The exact form of the cost function is nonlinear, but it can be approximated by a straight line, the tangent approximation in Figure 14.9. This permits the use of efficient linear programming or mixed integer programming methods.

When a linear programming method is applied the fixed costs cannot be taken into account. Therefore the linear underestimate of Figure 14.9 must be used. In

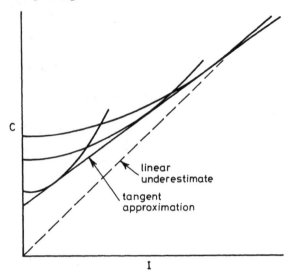

*Figure 14.9    Linear approximations for the cost function*

the mixed-integer-programming approaches the fixed costs can be included by applying 0–1 decision variables, which indicate the addition of a new substation or new feeder section.

Graph theory has proven to be very useful for the modelling of network planning problems. The distribution network is modelled as a flow network, where the directed arcs represent the power flow in feeder sections and the nodes are either source (substation) or load nodes (see Figure 14.10). The distribution-

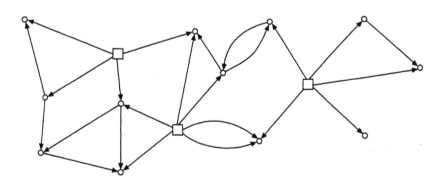

*Figure 14.10  Network modelling*

☐ substation node
o load node
↘ arc (power flow)

network planning problem can now be formulated as a minimum-cost flow problem.

For each arc in the network model a cost coefficient is defined, which indicates the unit costs due to the power flow through the arc, and a linear cost model is applied. If two or more arcs are connected in parallel and in the same direction, a piecewise linear approximation is achieved. Two parallel arcs with opposite directions indicate that two alternative directions are allowed for the power flow in the feeder section. For each arc the maximum capacity is also defined. It describes, for example, the thermal rating of the cables or lines.

For the solution of the minimum-cost flow problem many linear-programming algorithms are available. If the fixed costs are included the model applied is a fixed-charge network problem, for which the branch-and-bound algorithm can be applied. For the planning problem, nonlinear models have also been proposed. They give more accurate results but the computation time is longer. Usually conventional search methods are applied, but the use of genetic algorithms has been proposed as well.

The methods presented above have been developed for static planning tasks. In dynamic planning the timing of network investments must be determined. For dynamic planning a so-called 'pseudodynamic' approach can be applied. In a pseudodynamic approach a static planning model is sequentially applied for each state (year) of the planning period. The result can be enhanced by selecting the possible new facilities in a separate static optimisation procedure where the loads of the last year of the planning period are applied.

However, the pseudodynamic approach does not take into account the alternative investment sequences. The chain of investments covering the whole planning period can be optimised applying dynamic programming. The dynamic programming method efficiently solves problems formulated as multi-stage decision processes. The basic idea is the use of recursion: the optimal decision policy consists of optimal decisions. The optimal solution can be found by applying a recursion formula for each stage of the decision process. The number of comparisons is considerably reduced when each decision is studied separately instead of comparing all possible decision strategies.

For dynamic planning models, a good estimate of the future load growth is essential. However, there is always quite a large uncertainty associated with long-term load forecasts. The effect of this uncertainty can be taken into account by modelling the load growth as a Markov chain, and the resulting dynamic planning model is a discrete-time Markov decision process that can be solved applying the stochastic dynamic programming.

One of the drawbacks of sophisticated optimisation methods is the 'black box' effect: the planner has no idea how the computer reached the solution. This can be avoided by applying interactive planning where the planner can affect the solution process, e.g. by introducing case-specific constraints. Simple heuristic algorithms based on the expertise of planners can also be used. The creativity of the planner can be encouraged by letting him design the alternative network-development schemes, which are then put into the optimisation procedure.

*Network operation*

The network information system is connected to the SCADA system by a one-way link so that all changes in switch status and some measurement information transferred from SCADA are obtained in real time. By combining the SCADA and the AM/FM-GIS system a powerful environment has been developed for the real-time management of the distribution system. Some of these functions are introduced in the following.

*Switching state management:* The real-time switching state of the medium-voltage network is presented on the display of the workstation with the help of dynamic colouring which clearly distinguishes between individual feeder zones and de-energised line sections. A change in the status of any switch causes an immediate re-colouring of the network. The updating of the switching state can be made manually or originate from the information received from the SCADA.

*Switching schedules:* Switching schedules can be prepared in advance and stored in the database as chains of operations. A work order and a list of the customers who will be off supply is automatically printed for the schedule. Load-flow and fault calculations can be performed at any stage of the preparation.

*Fault location:* In the event of a short circuit, the system can estimate the likely fault location by using short-circuit calculations. Fault currents measured by protective relays are obtained through the SCADA connection. The estimated fault location is displayed and highlighted in a separate window.

*Trouble-call dispatching:* The receptionist receiving the trouble calls has access to the customer information. When customers are off supply they are indicated on screen. All calls are saved in the database for further analysis.

*Event log:* All events are saved in an event log which can be used for reporting and analysis. Outage times can be reported by customer category or component. Reasons for faults, and weather information can be stored for each fault.

*Applied computer technology*

The software of this network information system is based on GISbase® which contains the programmer's hardware-independent tools for the management of the database, user interfaces, graphics and output. Several operation systems are supported, e.g. OSF1, AIX, HP-UX and open/VMS. The architecture of GISbase utilises general standards and database management systems such as XII/Motif, Ingres and Oracle.

## 14.5.2  *Example 2*

This network information system has been developed by the Finnish company Versoft Oy, and is based on the research work carried out at Tampere University of Technology.

The general configuration of this application is shown in Figure 14.11. Only some of its special features such as the method for LV network design and MV reliability calculations are covered here.

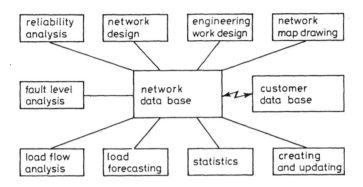

*Figure 14.11    Network information system: an example*

### *LV-network design*

In this application example, the use of personal computers for LV-system design is considered.

The LV-network design package to be introduced is primarily used to determine the MV/LV substation sites, and in the design of LV networks for new housing areas based on given town-plan development proposals. It can also be used for general network-reinforcement studies in both urban and rural areas. Using the system, the most economic solution is selected from those which are technically feasible, taking into account capital investment and system-loss costs.

The most important technical constraints are component thermal limits and network-voltage drops. Fault currents can be listed in order to check the operating times and discrimination of the various protective devices.

The optimisation program is based on a local-optimisation procedure. The designer can also specify different possible network configurations, so that their feasibility and costs can be assessed. The aim of the study is to find the optimum low-voltage network for the electricity supply to a particular area. By way of example, the substation sites and cable routes available are given in Figure 14.12a.

The solution procedure consists of three different parts. First, using all the preliminary selected substation sites, a network is found having a total line length that is as short as possible, but to which each disconnection box or customer is connected (Figure 14.12*b*). This is called the *minimal spanning tree* (MST). Instead of the line length, some weighted function, describing the relative excavation costs for a given area of ground, may be used.

The cable and transformer sizes for the MST network are determined on an economic basis. The first estimate of costs is calculated by adding the costs of losses to the investment costs.

A more economic solution may be reached by transferring the load flow to a suitable new branch, i.e. by adding one of the permissible line sections not utilised in the MST to the network. To retain a radial system some other line section or distribution substation must then be deleted. It is the planner's task to select the suitable branch to be added. The computer chooses the line section to be removed to give the most economical solution.

The total costs are calculated after each cycle of this local optimisation procedure. For each stage, the most appropriate economic cable sizes are used. If the total costs decrease, the new network image is adopted as the starting point for a new cycle of inserting a branch and opening a loop. When no further steps for reducing the costs can be found, this stage is terminated. In the simple example shown in Figure 14.12*c* the result includes only one substation. In general, the problems to be studied are more complex, and thus more transformers are needed.

There is no guarantee that this process will yield the global optimum. The intuition and experience of the planner are therefore important during the interactive procedure. Apart from this local optimisation procedure a 'forced changes action' has also been included. By using this procedure, the planner can form any kind of network configuration so that the computer can dimension the network and calculate the costs. Computer graphics is an invaluable aid in this kind of interactive design.

The third stage of the solution procedure is checking the voltage drop. If the voltage drops are too large, the most economical way of strengthening the network must be found so that it can satisfactorily supply the demands. Reinforcements are made in those line sections whose loads are only slightly below the economic limit, in which case the additional cost of moving to a larger cross-section is small. Other technical restrictions must also be checked if necessary.

The above program can also be used when considering, for example, how to reinforce an existing LV network. Here the alternatives are often either to uprate the existing lines or to divide the network into two or more parts, which then requires the construction of MV spurs and new substations.

LV-network design programs can be used successfully not only for standard types of schemes but also for developing recommendations and providing basic data for projects suitable for manual calculation. Another widely used application covers the provision of electricity supplies to meet local town or city development plans or the space-heating requirements of a new housing area. The program can investigate a number of options to derive the most economic method of providing these supplies based on different development scenarios.

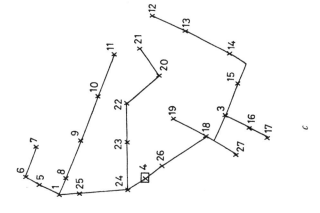

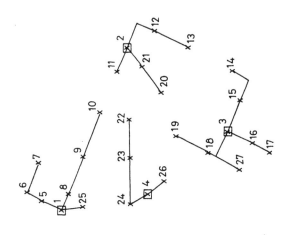

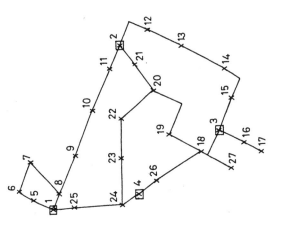

*Figure 14.12  Example of low-voltage network design*

STATION = METSAMAA

FEEDERS = NIEMI + LAHTI

OPEN ISOLATORS : 508 2031 399 397 2087 418 1111 414
TELECONTROLLED ISOLATORS: 419 420 410

RELIABILITY ANALYSIS

| SECTION | PROT. ZONE | DEMAND KW | AVERAGE OUTAGE RATES | | | | ANNUAL OUTAGE RATES | | | | COST OF OUTAGES | | | |
|---|---|---|---|---|---|---|---|---|---|---|---|---|---|---|
| | | | SUM 1/YR | FA 1/YR | MA 1/YR | TE 1/YR | SUM HRS | FA HRS | MA HRS | TE HRS | SUM £/YR | FA £/YR | MA £/YR | TE £/YR |
| 1  | 2658 | 2 | 80  | 10.2 | 3.3 | 0.4 | 6.5 | 2.8 | 2.2 | 0.4 | 0.2 | 322  | 227 | 19  | 76  |
| 2  | 408  | 2 | 94  | 10.8 | 3.3 | 1.0 | 6.5 | 3.8 | 2.6 | 1.0 | 0.2 | 344  | 231 | 40  | 72  |
| 3  | 407  | 2 | 55  | 10.3 | 3.3 | 0.5 | 6.5 | 3.2 | 2.5 | 0.5 | 0.2 | 210  | 148 | 15  | 47  |
| 4  | 448  | 2 | 66  | 10.5 | 3.3 | 0.6 | 6.5 | 3.5 | 2.6 | 0.6 | 0.2 | 179  | 124 | 14  | 40  |
| 5  | 406  | 2 | 281 | 10.8 | 3.3 | 1.0 | 6.5 | 4.2 | 2.9 | 1.0 | 0.2 | 1187 | 815 | 152 | 219 |
| 6  | 405  | 2 | 93  | 10.6 | 3.3 | 0.8 | 6.5 | 4.0 | 3.1 | 0.8 | 0.2 | 552  | 393 | 53  | 163 |
| 7  | 404  | 2 | 123 | 10.3 | 3.3 | 0.5 | 6.5 | 3.6 | 2.9 | 0.5 | 0.2 | 809  | 594 | 53  | 163 |
| 8  | 403  | 2 | 120 | 10.6 | 3.3 | 0.8 | 6.5 | 4.1 | 3.1 | 0.8 | 0.2 | 962  | 690 | 102 | 170 |
| 9  | 402  | 2 | 28  | 10.3 | 3.3 | 0.5 | 6.5 | 3.8 | 3.1 | 0.5 | 0.2 | 122  | 91  | 7   | 24  |
| 10 | 401  | 2 | 86  | 10.5 | 3.3 | 0.7 | 6.5 | 4.2 | 3.2 | 0.7 | 0.2 | 639  | 467 | 61  | 11  |
| 11 | 447  | 2 | 22  | 10.9 | 3.3 | 1.1 | 6.5 | 4.8 | 3.4 | 1.1 | 0.2 | 262  | 186 | 37  | 39  |
| 12 | 400  | 2 | 16  | 10.3 | 3.3 | 0.5 | 6.5 | 3.8 | 3.1 | 0.5 | 0.2 | 79   | 59  | 4   | 16  |
| 13 | 2696 | 2 | 32  | 10.7 | 3.3 | 0.9 | 6.5 | 4.5 | 3.4 | 0.9 | 0.2 | 222  | 161 | 25  | 36  |
| 14 | 355  | 2 | 43  | 11.0 | 3.3 | 1.2 | 6.5 | 4.9 | 3.5 | 1.2 | 0.2 | 264  | 187 | 34  | 43  |
| 15 | 6    | 3 | 123 | 13.7 | 4.2 | 1.0 | 8.5 | 2.8 | 1.5 | 1.0 | 0.3 | 845  | 495 | 107 | 242 |

| | | | | | | | | | | | | | | | |
|---|---|---|---|---|---|---|---|---|---|---|---|---|---|---|---|
| 16 | 421 | 3 | 38 | 13.4 | 4.2 | 0.7 | 8.5 | 2.5 | 1.5 | 0.7 | 0.3 | 136 | 82 | 12 | 43 |
| 17 | 2680 | 3 | 103 | 13.9 | 4.2 | 1.2 | 8.5 | 3.3 | 1.8 | 1.2 | 0.3 | 402 | 233 | 62 | 106 |
| 18 | 410 | 3 | 169 | 13.3 | 4.2 | 0.6 | 8.5 | 3.1 | 2.2 | 0.6 | 0.3 | 1144 | 768 | 100 | 277 |
| 19 | 226 | 3 | 52 | 13.3 | 4.2 | 0.6 | 8.5 | 3.3 | 2.3 | 0.6 | 0.3 | 221 | 146 | 17 | 58 |
| 20 | 354 | 3 | 6 | 13.6 | 4.2 | 0.9 | 8.5 | 3.7 | 2.5 | 0.9 | 0.3 | 23 | 15 | 2 | 6 |
| 21 | 409 | 3 | 18 | 14.0 | 4.2 | 1.3 | 8.5 | 4.4 | 2.9 | 1.3 | 0.3 | 101 | 66 | 13 | 23 |
| 22 | 398 | 3 | 2 | 13.5 | 4.2 | 0.8 | 8.5 | 3.8 | 2.8 | 0.8 | 0.3 | 8 | 6 | 1 | 2 |
| 23 | 1195 | 3 | 27 | 13.3 | 4.2 | 0.6 | 8.5 | 3.4 | 2.6 | 0.6 | 0.3 | 217 | 149 | 12 | 55 |
| 24 | 2093 | 3 | 9 | 13.1 | 4.2 | 0.4 | 8.5 | 3.3 | 2.6 | 0.4 | 0.3 | 41 | 28 | 1 | 11 |
| 25 | 2092 | 3 | 118 | 13.8 | 4.2 | 1.1 | 8.5 | 4.3 | 2.9 | 1.1 | 0.3 | 682 | 454 | 88 | 140 |
| 26 | 419 | 3 | 98 | 13.6 | 4.2 | 0.9 | 8.5 | 3.4 | 2.2 | 0.9 | 0.3 | 446 | 285 | 48 | 113 |
| 27 | 416 | 3 | 124 | 13.8 | 4.2 | 1.1 | 8.5 | 3.8 | 2.4 | 1.1 | 0.3 | 1172 | 755 | 129 | 287 |
| 28 | 417 | 3 | 12 | 14.3 | 4.2 | 1.6 | 8.5 | 4.6 | 2.7 | 1.6 | 0.3 | 49 | 31 | 7 | 12 |
| 29 | 1257 | 3 | 159 | 14.2 | 4.2 | 1.5 | 8.5 | 4.5 | 2.7 | 1.5 | 0.3 | 2514 | 1567 | 4078 | 541 |
| 30 | 415 | 3 | 108 | 13.6 | 4.2 | 0.9 | 8.5 | 3.7 | 2.5 | 0.9 | 0.3 | 664 | 435 | 65 | 164 |

COSTS OF THE ENERGY NOT SUPPLIED = 14818 £/YEAR

PERMANENT FAILURES = 9890 £/YEAR

MAINTENANCE OUTAGES = 1685 £/YEAR

TEMPORARY FAILURES = 3242 £/YEAR

AVERAGE DEMAND = 2304 kW

ENERGY NOT SUPPLIED = 8616 kWH/YEAR

FA = PERMANENT FAILURES

MA = MAINTENANCE OUTAGES

TE = TEMPORARY FAILURES

*Figure 14.13   Example results of reliability analysis*

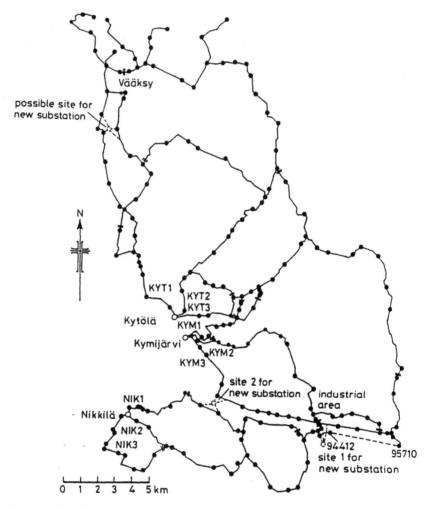

*Figure 14.14   Diagram of example network*

= open disconnectors
o primary substation
● distribution substation

The southern part of the network shown in Figure 14.14 covers the suburbs of a medium-sized town, while the northern area is rural. The city centre is not included in this survey. The system is fed by three 110/20 kV substations having a total capacity of 60 MVA, with an existing peak demand of 27 MW. The MV feeders are mainly overhead. It is required that various different possible developments of the MV system should be defined and compared, including any requirements for future new HV/MV substations. While overall there is load growth, the level of load varies considerably across the area being studied.

## 14.6.2 Present situation

With the existing peak load under normal feeding arrangements, voltage drops on the MV system do not exceed the accepted value of 5%, with the exception of the most northerly end of the network where drops of up to 6% can be met. There are a few sections of line totalling about 1 km, where, owing to economic reasons (high cost of losses), the existing conductors should be replaced by larger conductors. All lines can accept the passage of maximum fault currents during relay and circuit-breaker operating times.

When back-up feeding routes have to be used under outage conditions voltage-drop problems occur in the northern sections of the network. If any one of the feeders to the north is faulted, the maximum voltage drop rises to 11% and the associated underground cable at the source end of the feeder exceeds its thermal rating by 20%. In addition, loss of the southern HV/MV substation designated Nikkilä results in poor power supplies in the southerly area.

Studies carried out on the electrical condition of the existing MV network utilised the computer programs introduced in Sections 14.3 and 14.5.2, using the annual units supplied via each MV/LV distribution transformer to provide the load data.

## 14.6.3 Demand forecasts

Forecasts of annual load growth were based on area activity forecasts (population, jobs etc.) obtained from different local authorities. Owing to various uncertainties affecting load growth, scenarios for a low growth (L) and high growth (H) were produced in addition to the basic forecast (M). The results have been grouped together and converted to annual load-growth values (%) for each feeder, as shown in Table 14.2.

*Table 14.2  Estimated annual load growths (% /year) of MV feeders*

| Feeder | Years 1-5 | | | Years 6-10 | | | Years 11-15 | | |
|--------|---|---|---|---|---|---|---|---|---|
|        | L | M | H | L | M | H | L | M | H |
| NIK 1  | 4 | 6 | 8  | 2 | 4 | 6  | 1 | 2 | 3 |
| NIK 2  | 4 | 6 | 8  | 2 | 4 | 6  | 1 | 2 | 3 |
| NIK 3  | 4 | 6 | 8  | 2 | 4 | 6  | 1 | 2 | 3 |
| KYM 1  | 3 | 4 | 5  | 2 | 3 | 4  | 1 | 2 | 3 |
| KYM 2  | 6 | 10 | 13 | 5 | 8 | 10 | 3 | 4 | 6 |
| KYM 3  | 6 | 10 | 13 | 5 | 8 | 10 | 3 | 4 | 6 |
| KYT 1  | 6 | 9 | 12 | 4 | 6 | 8  | 2 | 4 | 5 |
| KYT 2  | 4 | 6 | 8  | 2 | 4 | 6  | 1 | 2 | 3 |
| KYT 3  | 6 | 8 | 10 | 3 | 5 | 7  | 2 | 3 | 4 |

L = low scenario; M = medium scenario; H = high scenario

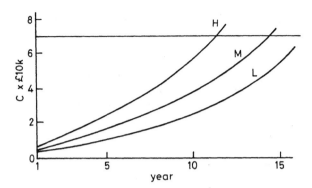

*Figure 14.15   Savings obtained from substation investment compared with the annuity of substation investment*

following options make it technically possible to delay construction for a few years.

- A contract could be negotiated with a neighbouring utility to purchase approximately 1 MW of demand, which is needed for the south-eastern corner of the supply area.
- A 4.5 km-long new feeder could be constructed between nodes 94412 and 95710 (Figure 14.14) and the feeding areas modified. Later this line could be used as a feeder from the new substation. The new substation could also provide additional transformer capacity to limit the loads on the present HV/MV substations, especially at Kytölä.

*Effects on the MV system*

The new substation on site 1 will offer the following benefits:

- The costs of losses and energy not supplied are reduced, depending on the load growth, by some £30 000 per annum at the beginning, the amount increasing annually by some 10%.
- Voltage drops are reduced.
- Earth-fault currents are reduced (the neutrals are isolated), and thus savings are made in providing earthing systems throughout the MV system.

The new substation will have the following disadvantages:

- High investment costs plus associated work such as new feeders.
- Increased fault level which leads to the need to replace the MV lines listed in Table 14.4, since the old small copper conductors would overheat in a 1 s 3-phase fault.

*Table 14.4  MV sections to be reinforced owing to increased fault levels*

| Line section | Length km | Type of existing line | Type of new line | Fault current, kA |
|---|---|---|---|---|
| 94461-93461 | 0·80 | Cu16 | ASCR99 | 5·5 |
| 94461-94360 | 1·01 | Cu16 | ASCR99 | 5·5 |
| 94360-97310 | 3·00 | Cu10 | ASCR99 | 4·0 |
| 94360-94310 | 0·50 | Cu10 | ASCR99 | 4·0 |
| 97310-97200 | 1·80 | Cu10 | ASCR63 | 2·4 |

*Table 14.5  Calculated feeder characteristics in year 15*

| Feeder | Peak demand, MW | Power losses, kW | Maximum load as ratio of thermal limit, % | Maximum voltage drop, % |
|---|---|---|---|---|
| NIK 1 | 6·1 | 28 | 35 | 1·0 |
| NIK 2 | 1·3 | 4 | 10 | 0·7 |
| NIK 3 | 5·8 | 41 | 74 | 1·6 |
| KYM 1 | 2·9 | 15 | 17 | 0·8 |
| KYM 2 | 2·5 | 13 | 32 | 0·8 |
| KYM 3 | 5·7 | 85 | 33 | 2·6 |
| NEW 1 | 2·0 | 10 | 15 | 0·9 |
| NEW 2 | 7·3 | 148 | 42 | 3·9 |
| NEW 3 | 5·7 | 65 | 52 | 1·9 |

Wherever economic, replacement of conductors should be carried out to obtain reduced losses. Table 14.5 identifies the most important electrical factors of each line following the replacements set out in Table 14.4. The parameters in Table 14.5 are based on the medium-growth scenario.

## Further studies

This example concentrated on the south-eastern corner of the supply area. The development of a reinforcement plan for other areas would follow the principles used here. The network-development options available for the northern area are construction of new feeders or a new substation. The former proved more economic if a low-load-growth scenario were applied, whereas the opposite applied in the latter case. Thus the recommended development plan would include new feeders at the start of the study period designed so that they could cope with the later construction of the substation. If a higher load growth should occur the policy would need modifying. The choice of constructing the new substation at the start would, however, be financially risky as far too much investment would have already been committed should the load growth turn out to be lower than forecast.

## 14.7 Bibliography

ADAMS, R.N., AFUSO, A., RODRIGUEZ, A., and GEREZ, V.: 'A methodology for distribution system planning. Proc. 8th Power Systems Computational Conference (PSCC), Helsinki University of Technology, Espoo, Finland, 1984

AGUET, M., FROMENTIN, M., CHARMOREL, P.-A., STALDER, M., PAROZ. G., and LEUBA, R.: 'Planning and operational systems of a city'. 9th International Conference on Electricity Distribution - CIRED 1987, AIM, Liege, paper a 02

AHLBOM, G., AXELSSON, B., BACKLUND, Y., BUBENKO, J., and TORAENG, G.: 'Practical application of computer-aided planning of a public distribution system' IEE conf. Publ. 250, 8th International Conference on Electricity Distribution–CIRED 1985, pp. 242-428

ALTONEN, J., AHONEN, A., KAUHANIEMI, K., and LAKERVI, E.: 'Experiences of applying stochastic dynamic programming for long-term of electricity distribution networks'. Proceedings of ISEDEM '93, Third International Symposium on Electricity Distribution and Energy Management, Singapore, 1993, Vol. 2, pp 695–700.

AUGUGLIARO, A., DUSONCHET, L., CATALIOTTI, V., IPPOLITO, M., and MORANA, G.: 'A package for the analysis of automated mv distribution networks'. Proceedings of Joint International Power Conference 'Athens Power Tech' APT '93, 1993, pp. 156–160

CASSEL, W.R.: 'Distribution management systems: functions and payback', *IEEE Trans.*, 1993, **PS–8**, pp. 796–801

CORY, B. J., 'Computer applications in power systems'. Proceedings of the first symposium on electric power systems in fast developing countries, Riyadh, 1987, pp. 34-42

COTTEN, W. L., MORISHIGE, N., and RAMSING, K. C. K.: 'PC-based design packages for electrical distribution systems', *IEEE Trans.*, 1991, **IA–27**, pp. 1141–1149

DAS, D., NAGI, H. S., and KOTHARI,D.P.: 'Novel method for solving radial distribution networks', *IEE Proc. Gener. Transm. Distrib.*, 1994, **141** (4) pp. 291–298

EFTHYMIADIS, A. E., HEATH, A. J. B., and YOUSSEF, R. D.: 'Interactive power system operations analysis with SCADA data capture'. Proceedings of Joint International Power Conference 'Athens Power Tech' APT '93 1993, pp. 367–371

FRANKEN, P.: 'The data bases of network-information-systems'. Distribution Conference 'Information Systems in Distribution' (UNIPEDE), Rome, paper 2.4

GÖNEN, T., and FOOTE, B.L.: 'Distribution-system planning using mixed-integer programming', 1981, 128 *IEE Proc. C*, **128**, pp. 70–79

GÖNEN, T., and RAMIREZ-ROSADO, I. J.: 'Review of distribution system planning models: a model for optimal multistage planning', 1986, 133 *IEE Proc. C*, **133**, pp 397-408

GÖNEN, T., and RAMIREZ-ROSADO, I. J.: 'Optimal multi-stage planning of power distribution systems', *IEEE Trans.*, 1987 **PWRD–2** (2), pp. 512–519

GROSS, G., and GALIANA, F. D.: 'Short-term load forecasting', *Proc. IEEE.*, 1987, **75** (12), pp. 1558–15737

'Introduction to integrated resource T & D planning, an ABB guidebook', ABB Power T & D Company, Raleigh NC, 1994, p. 599

KAUHANIEMI, K.: 'A probabilistic approach to the long-term planning of public electricity distribution networks', IEE Conf. Publ. 338, Third International Conference on Probabilistic Methods Applied to Electric Power Systems, PMAPS 91 1991, pp 311–316

KLEIN, L., KOGLIN, H.-J. FREUND, H., KÄMMERER, U., and WOLFF, H.-P.: 'Practical implication of interactive network planning', 9th International Conference on Electricity Distribution–CIRED 1987, AIM, Liege, paper 03

LAKERVI, E., JUUTI, P., and PARTANEN, J.: 'Aspects for electricty distribution network CAD-system developments'. IEE Conf. Publ. 250, 8th International Conference on Electricity Distribution–CIRED 1985, pp. 421-423

LAKERVI, E., KAUHANIEMI, K., and PARTANEN, J.: 'PC-applications for dynamic planning of network investments'. 13th International Conference on Electricity Distribution–CIRED, 1995, paper 6, 16, pp. 8–11

LAKERVI, E.: 'Aspects of electricity distribution network design and associated computer based tools'. Tampere University of Technology Publication 31, 1984

MICHON, R., FOURNIER, D., BOUQET, G., and PARANT, J. M.: 'PRAO - computer-aided network planning'. 9th International Conference on Electricity Distribution–CIRED 1987, AIM, Liege paper *a* 16

MIRANDA, V., RANITO, J.V., PROENCA, L.M.: 'Genetic algorithms in optimal multistage distribution network planning', *IEEE Trans.*, 1994, **PWRS–9**, (4) pp. 1927–1933

NORTHCOTE-GREEN, J. E. D., SILVA, R. F., DIAZ HERNANDO, J. A., and ADIEGO, J.: 'The use of computer aided distribution planning and design for the development of urban master plans'. 10th International Conference on Electricity Distribution–CIRED 1989, pp. 477–482

PARTANEN, J.: ' A PC-based information and design system for electricity distribution networks'. Tampere University of Technology, Publication 79, 1991

PARTANEN, J., JUUTI, P., and LAKERVI, E.: 'A PC based system for radial distribution network design'. Power Systems Computational Conference (PSCC), Lisbon, 1987

PARTON, K.C., EVANS, D., and THOMAS, C.P.: 'Computer aided design of distribution networks (CADARN)'. 9th International Conference on Electricity Distribution–CIRED 1987, AIM, Liege paper *a* 06

RAMIREZ-ROSADO, I. K., and GÖNEN, T.: 'Pseudodynamic planning for expansion of power distribution systems', *IEEE Trans.* 1991, **PWRS 1**, pp. 245–254

RENATO, C. G.: 'New method for the analysis of distribution networks', *IEEE Trans.*, 1990, **PWRD–5**, pp. 391–396

SUN, E. I., FARRIS, D. R., COTE, P. J., SCHOULTS, R. R., and CHEN, M. S.: 'Optimal distribution substation and primary feeder planning via the fixed charge network formulation'. *IEEE Trans.* 1982 , **PAS 101**, pp. 602-609

TERAZ, H. J., and QUINTANA, V. H.: 'Distribution system expansion planning models: an overview, Elec. Power Sys. Res. 1993 26 pp. 61–70

TRAM, H. ., and WALL, D. L.: 'Optimal conductor selection in planning radial distribution systems'. Westinghouse Electric Corporation, 1986

VAN SON, P. J. M., and CESPEDES, R. H.: 'Integration of real time control systems for electric utilities with emphasis on distribution management systems'. IEE Conf. Publ. 373. 12th International Conference on Electricity Distribution–CIRED, 1993, paper 4.18

WALKDEN, F. V.: 'Design of low voltage distributors', *Proc. IEE*, 1982 pp. 101-103

WALL, D. L., THOMPSON, G. L, and NORTHCOTE-GREEN, J. E. D.: 'An optimization model for planning radial distribution networks', *IEEE Trans.* 1979, **PAS–98** (3) pp. 1061–1065

WILLIS, H. L., and NORTHCOTE-GREEN, J. E. D.: 'Comparison of several computerized distribution planning methods', *IEEE Trans.* 1985, **PAS–104**, pp. 233–240

WILLIS, H.L., NORTHCOTE-GREEN, J.E.D., and TRAM, H.N.: 'Computerized distribution planning – data needs and results with incomplete data', *IEEE Trans.*, 1987, **PWRD 2**, pp. 1228-1235

below ground, but practices vary considerably. Nevertheless, in the interests of all parties, co-ordination of information is required and this will be considered further in the next section.

It is very important for an electricity-supply utility to have good contacts at all three official planning levels, as well as with individual private companies and contractors, so that the necessary background information relevant to future distribution developments is regularly updated. In practice, it is beneficial to everyone that regular meetings are held so that each side can understand the others' problems. The goodwill thus produced is invaluable in ensuring that the supply organisations have every opportunity of providing the most economic supply to any area, when required, whether in the centre of a large conurbation or in a small rural community. At all levels it is often necessary to work to prescribed formal arrangements, but the informal contacts built up as mentioned above can considerably ease the work load on municipal authorities, private companies and developers, and the supply utilities alike.

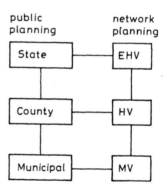

*Figure 15.1   Planning levels and their interactions*

## 15.3 Co-operation with other organisations

For its wellbeing a community needs to be provided with a number of service facilities — water, sewers, gas, steam or hot water, telephones and electricity. The first four are invariably located underground, and in built-up areas often all these service facilities have to be placed underground for environmental reasons. Given that each building requires a tapping from most, if not all, of these facilities, the ground underneath most roads and footpaths can contain a number of different services, especially where area trunk services follow the same route as the local services.

The large number of service facilities in built-up areas, particularly city centres, presents many potential problems wherever excavation takes place, for whatever reason. This requires the co-operation of the various service utilities to co-ordinate their work when having to maintain their own service facilities, in order to avoid damaging any of the other service facilities. In some cities large tunnels are constructed so that there is ready access to the various pipes and cables for repair and maintenance.

Various schemes have been developed whereby the service utilities interchange information on the routes of their services. Arising from this, the so-called 'one-call' system originated in the USA where service utilities provide one telephone number which contractors, local authorities, builders and the public can ring to obtain advice, information and supervision from the utilities concerned whenever any excavation is involved, and similar schemes now operate in many countries.

Where national maps are being digitised, this has provided a starting point for the service utilities to input data into a common databank. A method of interconnection between six utilities to a central databank is shown in Figure 15.2. Maps of different scales can be produced showing any combination of the utility services. For a particular work area, a large-scale map can be reproduced on, for example, A4-size paper for use by the work force excavating in that area, to avoid interfering with and damaging any of the services. An extract from such a map is shown in Figure 15.3.

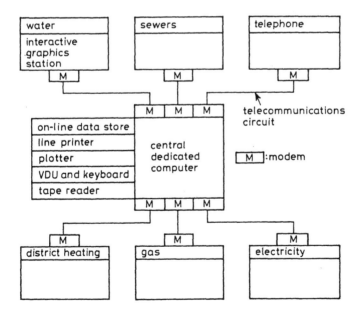

*Figure 15.2   Service utilities' co-operation: computer links to databank*

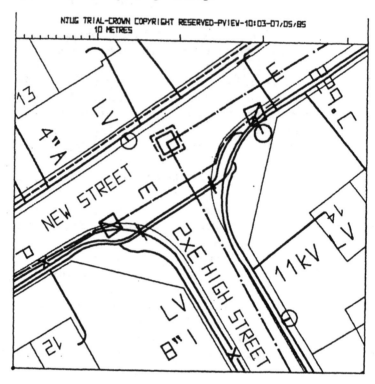

*Figure 15.3    Example of digitised map information (Courtesy National Joint Utility Group)*

Co-operation between neighbouring electricity-supply utilities ensures economic feeder arrangements and inter-utility network connections at the border areas, to the benefit of each utility. In addition, it may be more economic to receive back-up supplies under outage conditions from a neighbouring utility at agreed cross-border energy-tariff rates, rather than reinforce the tail end of one utility's network. A common pool of such items as HV/MV transformers and special maintenance equipment or vehicles can avoid money being tied up in idle equipment. The joint use of training centres and exhibitions, and co-ordinated approaches to provincial and national level administrations, can cut costs and ensure more consistent policies that are likely to be accepted by other organisations. The use of common research centres, equipment testing stations, and working groups to solve mutual problems cuts down on the amount of overlapping work and reduces costs still further. There may also be advantages on occasion for a number of utilities to use a common control centre.

At HV the smaller utility may find it worthwhile utilising the expertise and resources of a larger power board in planning, designing, constructing and

*Table 15.1  Contributions and demands on a utility*

| Group | Contribution | Demands |
|---|---|---|
| Owners | Own capital, risk infrastructure for successful business | Return on investment, continuity, influence on society, good image |
| Customers | Buy energy, pay service charges. Appreciation and acceptance of supply | Reasonable price, high quality of supply, good level of service |
| Society | Legislation organisational stability | Stable power-supply taxes, low environmental risks, safety, employment |
| Financiers | Capital from outside the utility | Interest, safety of investment |
| Management | Organisational skills | Salary, status, managerial freedom |
| Staff | Work input (skills, techniques, knowledge) co-operation | Salary, status, safety, continuity of employment, working conditions |
| Manufacturers and contractors | Provide satisfactory appliances and service, good co-operation | Exchange of information |
| Other distribution companies etc. | Co-operation, specialist knowledge, development work, consultancy and information service | Resources, joint interest in development projects, loyalty, exchange information |
| Testing laboratories and research centres | Quality control, research results | Input material, financial aid |
| Universities and other schools | Education and research publications | Training, help with financing, information exchange |

indicates land protected under state or local-authority regulations, including Green Belt areas around conurbations and nature reserves.

Within these areas sometimes the only way of obtaining permission from planning authorities for a route for electricity supplies is to underground a section, even at high voltage, notwithstanding the considerable increase in costs.

When planning authorities request that a proposed new HV line should be underground in a particular area, where MV and LV overhead lines exist already, it may be possible to reach an agreement that these MV and LV lines are undergrounded, more cheaply than HV lines, in order to obtain a right of way for the new HV line.

Turning to MV/LV substations, for economic reasons open ground-mounted distribution-transformer substations are used, even in built-up areas, wherever possible. However, the higher land costs, the development of smaller prefabricated MV/LV substations, plus pressure from municipal planning authorities is leading towards distribution substations being installed within prefabricated housings made of suitable materials to blend into the surroundings. One advantage is that, being indoors, the equipment is then not so much at risk from weather conditions or vandalism. Suitable arrangements must be made to avoid any oil leakage from transformers from escaping from the substation site and causing pollution problems. Costs can sometimes be reduced by converting suitable unused buildings like chapels, churches or warehouses. Larger outdoor substations can be concealed to varying degrees by earth embankments and tree plantations.

The passage of 50 or 60 Hz current through transformers produces mechanical vibrations, caused by magnetostrictive forces in the core resulting in a noise of 100 or 120 Hz being emitted from the transformer. In most situations this small noise is no problem but, especially in cellar installations and sometimes in quiet residential areas, and in the countryside, it can lead to complaints from nearby residents. The problem can sometimes be overcome by changing the unit for a less noisy one, or by installing vibration dampers. Planting of trees around a site reduces the noise level slightly but, for larger substations, it may be necessary to surround the transformers by a brick or concrete wall. Where HV/MV transformers are situated in close proximity to houses it is invariably necessary to install them with noise-reducing enclosures surrounding the transformer tank. The dimensioning of the enclosure, its distance from the tank, and the distance between the inner and outer linings are critical for good noise reduction. Transformer noise can also be transmitted through the ground, dependent upon the local geological conditions.

There has been much discussion and considerable research into any possible health effects from 50—60 Hz electric and magnetic fields. Electric fields are weak owing to the relatively low voltage levels in distribution systems and, in general, magnetic fields associated with MV networks are also low because the 3-phase system tends to balance out the fields from separate conductors. In addition, the normal clearances between a line and an individual person attenuate the field and thus reduce any risk. Although any risks seem to be quite low, some prevention of continuous predisposition to fields above 0·25 or 1 μT is recommended. In public electricity distribution networks the most relevant object for this consideration is the low-voltage busbar of a basement distribution substation. Usually a 4 m distance from the low-voltage terminals to the nearest dwelling room is satisfactory. Alternative means of decreasing the magnetic field would be to shorten the busbars or replace them by a cable.

With the older established lines, at all voltages, housing and industrial development has now surrounded the lines in many areas so that access to the conductors and towers for maintenance and repair can be a considerable problem. Similarly, precautions have to be taken so that access to any substation will not be prevented by local residents, for instance by parked cars. Within urban areas space requirements have been drastically reduced by the use of gas insulated switchgear.

All these amenity aspects place additional costs on a utility. If the requirements are imposed by government, county or municipal authorities then compensation is sometimes possible. However if an MV line were under-grounded to permit an area of land to be developed for industrial or housing purposes, the developer would often be expected to pay all, or part, of the cost involved.

## 15.5 Manufacturers and consultants

Power-distribution utilities often have their own standards and specifications for equipment, and construction practices which may be more strict than, or equal to, national or international specifications. It is essential that a manufacturer is fully aware of a given utility's codes of practice in the planning, design and operational areas in order to produce a suitable item of equipment. Frank and honest discussions, and the acceptance of prototypes of new equipment at installations on electrical networks, can promote such developments. When deciding on manufacturing standards, or engineering recommendations, a utility should keep in mind the opportunity of using a sufficient number of manufacturers to obtain competitive quotations and a guaranteed continuity of supply of equipment for the future.

Manufacturers will naturally provide advice on their own products and systems. Independent consultancy advice is often useful for new systems or methods, or those based on foreign practices, which may not be familiar to the utility personnel. The use of consultants is also common to provide general advice on various aspects when any electricity-supply scheme or utility is just starting up. In developed countries small utilities tend to use consultants for new items such as the introduction of $SF_6$ switchgear, or the purchase of a major telecontrol system. Equally, a large distribution utility may use the services of a consultancy firm for a temporary period. This could be to overcome a short-term peak load in work on the planning and design staff. In addition, a general plan for system development by an outside consultant can often bring a new viewpoint to the problem to be discussed with the utility, or reveal errors in the utility's existing planning practice.

## 15.6 Universities and research centres

In many countries it is common practice for university staff to be invited by utilities to attend 'brain storming' discussions on specific problems, with variable success. Nowadays, in addition to basic research, universities are often capable of carrying out research programmes for utilities on a contract basis. These may deal with new ideas for network automation, the production of suitable CAD packages for network planning and design, and designing an analytical model for equipment such as an induction motor or inverter/convertor installation. Larger projects are often jointly financed from a suitable national fund, in some cases government-based. Post-graduate and career-training courses suitable for practising engineers are also available to upgrade engineers educationally, and/ or update their technical know-how. The participation of utility staff not only extends their own capabilities, but also provides a fund of practical engineering experience of information on existing distribution systems, and the problems encountered, for the university faculty members and students.

Unlike research centres or commercial consultants the universities do not usually have permanent personnel for contracts financed by the government. Nevertheless, those universities which undertake contracts and have good liaison with supply utilities usually provide teams for research. They can also provide well structured courses, so that their graduates are more likely to obtain worthwhile jobs with utilities.

Some research centres have independent personnel and specialised equipment enabling them to test new types of equipment being considered for purchase by utilities. In addition to testing to generally accepted standards, they are also able to perform customer-specified tests and obtain information on particular characteristics of the equipment under test. This is of increasing importance in open market conditions, and when utilities are looking towards extending the useful life of their equipment.

## 15.7 Bibliography

AHRENT, H., and AMBROSCH, H.: 'Ästhetische Beeinflussung der Umwelt durch Hochspannungs-Freiluftschaltanlagen (Aesthetic influencing of the enviornment by high voltage outdoor substations)', *Elektrie*, 1983, **37**, pp. 341–345

DELCAMBRE F.: 'Acoustic protection of the environment for transmission and distribution networks', *Rev. Fr. Electr.*, 1982, **55**, pp. 4–11

FOSTER, S.L., and REIPLINGER, E.: 'Characteristics of, and control of, transformer sound', *IEEE Trans.*, 1981, **PAS-100**, pp. 1072–1077

JOUNIOT, J.A., and LACOSTE, J.: 'Medium and low-voltage overhead lines and the environment'. IEE Conf. Publ. 250, 8th International Conference on Electricity Distribution — CIRED 1985, pp. 180–184

KANNUS, K., LEHTIO, A., and LAKERVI, E.: 'Radio and TV interference caused by public 24 kV distribution networks', *IEEE Trans.*, 1991, **PWRD–6** (4), pp. 1856–1861

KENNEDY, M.W.: 'The consulting engineer — a catalyst for progress', *IEE Proc. C*, 1987, **134** pp. 97–103

LAKERVI, E., and PARTANEN, J.: 'An attractive curriculum in power engineering', *IEEE Trans.*, 1992, **PWRS–7**, pp. 346–350

LAKERVI, E., and PARTANEN, J.: 'Experiences of combined university courses for students and engineers in electrical power engineering'. SEFI Annual Conference on Engineering Education in Europe, Louvain, Belgium, 1988 paper W2-2

PERRY, T.S.: 'Today's view of magnetic fields', *IEEE Spectr.*, Devember 1994, pp. 14–23

PESQUI, R.: 'Electricity distribution and the environment', *Rev. Fr. Electr.*, 1977, **50**, (259), pp. 42–47

VIS, E. W.; 'Transformer noise in residential areas', IEE Conf. Publ. 151. International Conference on Electricity Distribution — CIRED 1977, pp. 80–83

YARROW, G.J.: 'Digital records trial — NJUG Dudley', *Distrib. Dev.* March 1983, pp. 6–9

Printed in the USA
CPSIA information can be obtained
at www.ICGtesting.com
JSHW011518221024
72172JS00008B/61

9 780863 413094